SMILING SECURITY

SMILING SECURITY

THE CYBER SECURITY MANAGER'S ROAD TO SUCCESS

MIKKO NIEMELÄ
PASI KOISTINEN

LIONCREST

PUBLISHING

SMILING SECURITY

The Cybersecurity Manager's Road to Success

ISBN   978-1-5445-1179-5 *Paperback*
        978-1-5445-1180-1 *Ebook*

CONTENTS

FOREWORD

So, you've decided to pick up a book about cybersecurity? If you're not a decorated warrior from the tinfoil hat brigade, that probably means that you're working in either the IT, risk or governance fields. Or at least you're interested in one of those areas, whether you knew it or not.

In a world of exponentially increasing data collection and retention, organisations and individuals are creating a spectacular catalyst for future disaster. Yet, almost invariably, that potential liability remains steadfastly hidden behind the semi-technical belief that 'someone from IT looks after that stuff'.

There are a multitude of arguments to be made about how to value business investment in IT capability. Regardless, the rational premise for such investment is that, in doing so, there is a productivity or efficiency gain. What might be called, in military doctrine, a force multiplier. Give three of your brightest minds a calculator each, or introduce them to Python? You know which is going to create greater value instinctively. Of course, the logical counter-corollary to that instinct is that anytime you observe an individual with a spreadsheet open, and yet they are using a calculator, you've just identified someone from outside the Venn diagram containing the three bright minds.

One way or another, IT capability is about using, processing or manipulating data. It might be real-time data like the packet stream in VoIP. Or it might be at the other end of the scale, where every possible piece of uncorrelated information is unceremoniously dumped into some enormous Hadoop-enabled data lake/pool/warehouse/crypt for later investigation. The latter might not be intended for the purpose, but it will eventually and convincingly become evidentiary proof that correlation does not imply causality.

It's quite normal for the productivity and efficiency gains from investment in IT capability to be accounted for prospectively in some business case made by clever people from the technical realm. It's way less likely that those same folks will put a potential downside figure on the damage that might be done should the clever tech solution be breached and the data leak into the wild. And when that breach happens, as it surely will (or possibly has already, and you don't know about it just yet), who's responsible? If you're in a big enough organisation to find yourself publicly listed, you'll probably have a CIO. Think that she's responsible? Check the corporation's legislation in your vicinity. Bet you don't find mention of the CIO; CEO possibly, president maybe, chairman likely, directors definitely.

If you're running a smaller organisation and don't have the pesky corporation regulators leaning over your shoulder, don't worry, the responsibility is probably yours directly. Even if the eleven-secret-herbs-and-spices recipe of your business success leaking into the big wide world doesn't bother you immediately, the liability arising from leaking private client information within your control might trigger the local privacy commissar to drop in. The belated realisation of the Fakebook generation that having all of your freely but ignorantly given personal data and intimate details of your online behaviour and thoughts being collected and auctioned for corporate gain is that it feels kinda bad. So comes the latch for the privacy gate well after the data horses have long since bolted. Privacy tsars have punitive legislation at their disposal that makes corporate regulators turn Hulk-green with jealous rage.

So why do I tell you this? How about a purely hypothetical case study; what Schrödinger might have described as a thought experiment, albeit one without any feline mortality.

Imagine that you've just taken over the IT department of a twenty-year-old e-commerce company with an annual turnover of $1 billion and the strongest brand recognition in your country. Now imagine that after you took over, you found a few things that started to make you feel a little uncomfortable about the resilience of your organisation. DR sites that had never been tested. One admin password for all functions that hadn't been changed in eleven years! Backup scheduling that had no apparent risk/reward correlation, and that's for those systems where a backup actually existed.

Sometimes it's better to be lucky than good. However, having been long involved in risk management, I continue to believe that hope is not a strategy. Not a good one at least. Certainly not one that will satisfy any of the myriad investigators when they apply their retrospectoscope to your next big breach.

Strangely, it was after Mikko and his guys had performed some rather enlightening (and frankly, quite horrifying) penetration testing that the first privacy breach occurred. And typically, it wasn't from any obvious external source but from an internal control failure which saw a patch promoted to production that had already been identified as flawed and not fit for release.

The potential release of privacy-protected data was relatively small in size and severity, particularly compared to some of the more spectacular and complete recent failures from such data giants as two major airlines, (anti-?) social media companies, dating sites, financial institutions, government health departments, government security verification providers (?), amongst many others. Self-declaration to the local privacy Gruppen-führer and internal audit sniffer-dog team led to some moments of soul-searching and nervous dusting off and rewriting of curricula vitae.

However, the final jewel in the crown, the keystone in the Arche de la Defence, the cherry on the cake that contained the story of this being a small and minor breach was the satisfaction of showing that we had breach monitoring in place through Cyber Intelligence House. Their monitoring showed that none of our potentially leaked data had appeared in any of the malicious sites where such data would appear. Case closed.

All hypothetically, of course.

The insurance industry is offering more and more options for insuring against cyber risk. They're not doing so because they're in need of more opportunities to hang out with the folks from IT departments around the corporate world. They're doing so because more organisations, their directors and executives are becoming aware of the potentially massive liability that comes from the new world of data everywhere. There's a gaping chasm between the people who will likely seek this book out and those who really need to. If data is the new oil, don't become famous as the captain of the first digital equivalent of the *Exxon Valdez*!

—RICK HOWELL, AIRLINE EXECUTIVE

INTRODUCTION

PLEASE PUT OUT OUR FIRE!

When is the best time to buy fire insurance: before or after your house burns down? That's a rhetorical question. The answer is obvious.

By the same logic, when is the best time to invest in cybersecurity: before or after your company gets hacked?

Unfortunately, most companies and organisations invest in cybersecurity or hire a cybersecurity manager only *after* they've experienced a data breach and seen the tremendous damage caused by the breach. In other words, by the time they take action, the damage is already done. That's like trying to buy homeowner's insurance after the house is already burning down.

The unfortunate fact is that most organisations do not invest in cybersecurity before they get hacked. They only take it seriously after the fact. Why? Because business leaders who have no experience with cyberattacks mistake it as a low priority. They think of security measures as overhead costs, so they don't hire a cybersecurity manager (CSM) until they absolutely have to. When they do hire a CSM to solve their problems, they don't allocate any resources beyond the CSM's salary. That means the CSM is walking into a fire with no tools to put out the flames and prevent future fires.

We see it happen all the time. Despite the high risk and incredible costs of a cyberattack, business leaders are simply reluctant to spend money on cybersecurity. Recently, we met with a company whose entire email system was hacked. This disaster threatened the whole operation, but when we asked them if they were willing to spend money to fix the problem, they said, 'No, we don't have any budget for that.' Less than a week later, we met with a company that had just lost $400,000 in a cyberattack. We had given them a proposal for a solution that cost $20,000. They had said it was too much.

A COSTLY EPIDEMIC

Is it too much? Not by a long shot.

According to one study by Kaspersky Lab, the average cost of a data breach in the United States is $1.3 million for large businesses, and more than $100,000 for small and medium-sized companies.[1] More and more high-profile companies are being breached all the time because hackers are becoming more sophisticated.

Over the past several years, we've watched the problem grow to unbelievable proportions. A few examples: Yahoo!, three billion user accounts breached. Adult FriendFinder, 400 million user accounts. eBay, 145 million user accounts compromised. Equifax, 140 million. Target stores, 110 million. JPMorgan Chase, 76 million. Anthem Health, 78 million. Home Depot, 56 million. Adobe software, 38 million. Most people in the world have been affected because almost every person uses the internet, and people are what makes companies run.

BACKWARD THINKING

Many clients who call us say things like 'We knew this security stuff should have been done long ago. We knew we were vulnerable, and we had weak passwords.' In other words, they ignored the risks. By procrastinating or disregarding cybersecurity, companies are taking a

---

[1] "Kaspersky Lab Survey: Cyberattacks Cost Large Businesses in North America an Average of $1.3M," Kaspersky Lab, September 19, 2017, https://usa.kaspersky.com/about/press-releases/2017_kaspersky-lab-survey-cost-of-cyberattacks-for-large-businesses-in-north-america.

tremendous chance. We call this taking on *cybersecurity debt*. The only question becomes 'When is that debt going to have to be paid?'

Failing to prepare for these attacks is shortsighted. Basic cybersecurity is actually not very difficult to put in place if the company takes steps before an attack happens. Most companies can at least protect the low-hanging fruits that are at greatest risk, like employee records that contain identifying information.

That said, even the best protected can fall victim to cyberattacks. Attackers are agnostic—if they find compromised accounts anywhere, they'll attempt to breach them. That's why companies need to do more than they've been doing. They need to think far outside the company's reach to the ways their employees expose their identities online.

Most of the time when companies do decide to hire a cybersecurity manager, they aren't thinking that broadly. They do it to put out a fire. The CSM's mission is disaster recovery with the immediate objectives—and sometimes the *only* objectives—of controlling losses and mitigating damage. In these cases, the CSM will have to work for months just to solve the current crisis. Then, maybe next year, they will have time to start building a defensive, preemptive cybersecurity strategy.

COMMON TYPES OF CYBERATTACKS

There are many types of cyberattacks, but the vast majority of the malicious hacking of businesses is conducted by organised crime rings with only one motive—profit. These hackers steal private information for corporate espionage. Most fall into four broad groups.

The first group is made up of hackers we call 'hacktivists'. In the name of a social-justice cause, hacktivists may deface a website or leak classified or private data. They aim to damage an organisation or harm certain individuals to make a political point. Hacktivists make up a relatively low percentage of cyberattacks.

The second type of attack is known as 'ransomware'. With ransomware, the hackers don't even need to transfer any data out of the company; they just need to encrypt it so it cannot be used. That's enough to bring many companies to their knees. Most victims pay the ransom. They just want their data back, whatever the cost.

A third type of cyberattack is a bit different from most—it's executed in the name of an ideology, not necessarily for money. These hackers' only goal is to inflict as much damage as possible. Ideological attacks may come from disgruntled employees or people who are mentally unstable. For example, one fired system administrator got revenge on his company by encrypting and changing everyone's passwords before he left.

The fourth major group of cyberattacks is carried out by nation-states, usually through their intelligence agencies. Their goal is to steal state secrets and classified documents. This type of attack has become much more frequent over the past five years. Hackers get into networks and collect information on other nations, with the intent that it will be used in hostile actions between countries. These attackers have become quite sophisticated and often pull off their schemes anonymously.

When most people think of countries responsible for nation-state attacks, they think of China, but it's a public secret that everybody is doing these attacks now. Most of the world tries to keep pace; it's like the Cold War, but with an incredible reach and billion-dollar budgets. It's definitely a global problem; consider the Wassenaar Arrangement, signed by most countries in 2017, which effectively placed cyberweapons under the control of an international arms agreement. Tools for professionals have now been weaponised.

We think that's backwards. Companies should hire a CSM *before* they suffer a cyberattack. As the saying goes, an ounce of prevention is worth a pound of cure. It's far more effective to proactively build a defence against cyberattacks than to respond to an attack after it happens.

If an organisation wants to be safe from cyberattacks, they have to care about security. They have to prioritise it. They must realise that it's not a discretionary expense. This means hiring a cybersecurity manager and then allocating budget funds to pay for what needs to be done.

If an organisation wants to be safe from cyberattacks, it has to care about security.

CYBERSECURITY: A PEOPLE PROBLEM

Because cybersecurity risks are so huge, companies need to recognise the high value the CSM brings to the organisation and hire excellent CSMs. Most do not. Why? Perhaps because the CSM's role can be very low-profile, even invisible to many people in the company. Yet the CSM can create huge wins for the company—and perhaps even more important, they can prevent the company from suffering enormous losses. They are agents of change who have the power to transform companies for the better; they should always be treated with respect and gratitude.

That mind shift happens when companies learn to treat cybersecurity not as a technical problem but as a people problem. The basic strategies of cybersecurity haven't actually changed much over the past twenty years. In the year 2000, the most common tactic used to infiltrate a company's IT network was to send fraudulent phishing emails to employees. Or hackers might have taken advantage of bad passwords used across different internal and external services. Or maybe they would find company servers or workstations that were rarely updated and hack those.

Nearly two decades later, many cyberattacks are exactly the same: they're low-tech and are set in motion by a distracted employee clicking on the wrong link in a suspicious email or an entire sales team using the same simplistic login and password—such as *admin* and *admin*—to access the company's computer systems.

What's changed is how effective these attacks are. People use so many credentials all across the internet that they have a hard time remembering them all, so they use the same username and password everywhere. All a hacker has to do is find a username (usually an email address) and they're in.

So we can see that security isn't about computers, servers, firewalls and software. It's about human behaviour. It's about the passwords people choose. It's about their use of Facebook, Skype, Dropbox, cloud services and so much more. Controlling human behaviour is a big part of the CSM's job.

The ideal hire for the CSM role is not a techy geek who is antisocial and plays with computers all day. Computer skills are only a small part of the job. The best CSMs can understand what ransomware is and research the details themselves, comprehend high-level business strategy and risks, communicate effectively inside the organisation, demonstrate solid interpersonal skills, work with a budget, complete projects, work as a team member and understand human behaviour and project management.

ABOUT THIS BOOK

Cybersecurity is not a particularly fun or funny topic. We know that. Cyberattacks can be devastating, and they can ruin companies and careers. But too many people incorrectly think of cybersecurity as necessarily difficult, frustrating and a nuisance. Sort of like the airport security process, they expect it to be uncomfortable, not a positive or feel-good experience. We want to change that perception.

In this book, we present an approach that we hope will encourage people to view security in a positive light and see it as a worthwhile and beneficial element of the company. When managed properly and proactively, cybersecurity can reduce worry, eliminate stress and increase confidence. Sound security also allows companies to boost profits and revenues by taking more risks in business, without the fear of an attack. That's why we titled this book *Smiling Security*.

At least two types of readers will find valuable information in this book. The first is any business leader who is hiring, or knows they should hire, a cybersecurity manager. Millions of companies—small, medium and large—get the hiring part wrong. Maybe they have never hired a CSM before, so they recruit the wrong skillset. Or they don't know what it takes to be a competent CSM, so they end up hiring the wrong person. As a result, they get inadequate results.

CSMs themselves will benefit equally from the information we share here. Being hired into a company as a CSM can be challenging. The CSM must learn a tremendous amount of information in a short period of time, while also navigating the corporate culture of the company and identifying vulnerabilities. We'll explore these challenges throughout the book.

This book will help organisations of all sizes appreciate the role of the cybersecurity manager and the value they can bring to a company. By the end, no matter what your role in an organisation, you will understand how cybersecurity works and how the cybersecurity manager fits within the organisation. The knowledge held in these pages will help organisations become stronger, safer, and less likely to suffer a cyberattack. It will also help CSMs better understand their role within an organisation and show them how to make the greatest impact. To that end, we will split the book's content into three parts:

- Part I: Discovery: Understanding the Company
- Part II: Communication: Working with Each Team to Create Change
- Part III: Process: Securing Eight Domains in Ninety Days

By providing a roadmap for cybersecurity, the book will help organisations successfully build and operate a cybersecurity department from the ground up, effectively securing the organisation in the shortest amount of time possible. In twenty-two chapters, we will examine the necessary convergence of security and business administration, communication and catalysing change.

ABOUT THE AUTHORS

We have a combined thirty-plus years working in all aspects of cybersecurity. Currently we own two cybersecurity companies. One is a cybersecurity consultancy; our customers hire us to provide ethical hackers to perform cybersecurity testing services. In other words, we challenge their cyber vulnerabilities.

At Cyber Intelligence House and Silverskin Information Security, we do cybersecurity and exposure assessments and ratings. Similar to a credit rating company that rates the repayment risk of bonds for investors, we provide ratings for a company's cybersecurity risk profile. We help to create certainty in a situation that is inherently uncertain. Most people don't know what to do about cybersecurity. They don't know what to worry about and what to ignore.

We help our clients find and hire qualified CSMs that best fit their needs. We also help CSMs solve their problems. And we provide CSM services to companies that don't want to hire a full-time CSM position.

We work with medium and large companies across a wide range of industries in both the public and private sector, including government agencies and nonprofits. So we have broad and deep experience in the latest cyber threats and state-of-the-art security. When it comes to implementing effective cybersecurity, we've seen it all, and we know every aspect of the industry. We're going to share that information in this book.

PART I

DISCOVERY

UNDERSTANDING THE COMPANY

A successful cybersecurity manager must understand the company they're working for. They have to know how the organisation is structured, who makes the decisions, who plays what roles in each department and how the employees experience and demonstrate the company culture.

CSMs will need internal and external help to accomplish their objective. They will also need resources, so they must understand who can approve budgets and how those decisions are made. It's essential for the CSM to get to know the company leaders and their expectations and motivations, as well as the company's key personnel and stakeholders, current liabilities, plan, business environment and budget.

Who are the key players in the company? The CSM must determine which people support their mission and which ones might oppose it, then identify who is amenable to change and who is not. Change will be required, so identifying obstacles to change early on is crucial.

In part I, we will explore strategies for gathering and using this information so the CSM is prepared to navigate their new environment with maximum efficiency and effectiveness in executing a formal cybersecurity plan.

CHAPTER ONE

KNOW THE EXPECTATIONS (OR LACK THEREOF)

The ideal situation: a newly hired CSM reports for duty and, on his or her first day, is presented with a set of specific expectations and detailed objectives for cybersecurity set forth in a clearly written cybersecurity plan. The plan provides the CSM with concrete goals to work toward. It allows the CSM to hit the ground running and achieve maximum results in the shortest time possible. The CSM will be extremely effective because the company has a detailed cybersecurity plan in place.

Sound good? When it happens, it is. But how often does it happen? Almost never.

Most of the time, the company has no plan, no set goals, no idea of scope and no clear expectations for the CSM they just hired. Typically, the company is only able to express to the CSM a vague desire, like 'We want our data to be safe' or 'We don't want to get hacked again.' That's not a plan.

In fact, most of the time, the company expects the CSM to come up with the plan. They also expect the CSM to resource, scope and execute the plan. The CSM, meanwhile, depends on the company to define parameters and provide a budget. With each party operating under the impression that the other should be providing the basic necessities, the CSM's role often stalls out within the first week or two on the job.

It's not that the company wants to leave the CSM in the lurch. Quite often, when a CSM is hired, it's the first time the company has ever hired a dedicated cybersecurity manager. The company has no experience with having a CSM on staff. They have no history or established protocol for how to manage a CSM. They have no idea how to best utilise this new asset. The company isn't thinking about what additional resources or support the CSM may require; they are primarily focused on budget.

HIRING A CSM: WHAT TO ASK IN AN INTERVIEW

- How did you build and manage security management systems in previous jobs?

- How would you set one up here?

- What do you think are the most relevant cyber risks for businesses like ours?

By listening to the candidates' answers, you'll find out if they are up to date on current risks and if they did their homework about your company. Their answers will also reveal the mental boundaries that limit their work.

MISMATCHED TALENT

When a CSM is hired, they're usually expected to function independently of any one department or team. Within any company, there are internal teams, or tribes, such as operations, finance, sales, marketing, legal, IT, brand management and so on. Since the cybersecurity role lies outside those power teams, the CSM has limited influence within the company. The way forward for the CSM is to work within the internal corporate structure to get on the agendas of those teams and personally influence the key stakeholders and decision-makers.

In many ways, the CSM's effectiveness is limited by how well they can navigate the internal power structure of the company. CSMs have to work hard to gain acceptance into these tribes. This part of the job is something that is never mentioned in working contracts, it's seldom taught in universities and it can come as a surprise for CSMs without significant experience.

For example, we know of a company that hired a very proficient-looking CSM with an impressive CV. While the CSM had deep technical experience, he lacked the people skills to work effectively within the company's corporate culture. Instead of taking the initiative and proactively forging relationships, he waited around for department heads to invite him to a meeting. He ended up sitting in his office all day on the computer instead of communicating with company leaders and managers. In the end, he achieved little to improve the company's cybersecurity, all because there was a mismatch between the company's expectations and the CSM's expectations. Each was waiting for the other to take action.

Sometimes companies don't even know what kinds of skills the cybersecurity manager should have, so they end up hiring the wrong kind of talent with the wrong expertise. Hiring a cybersecurity manager with the wrong skillset is going to end in failure because the person hired doesn't match up with the actual needs of the company. But it's not the CSM's fault, because the company didn't know what they needed to begin with and didn't make it clear during the hiring process.

If the company misunderstands the role and hires the wrong person, then little will be accomplished.

PROBLEMS LIE AHEAD

Most companies hesitate to commit sufficient resources and budget for a robust cybersecurity programme. They often believe that the cost of hiring a CSM is the only investment they will have to make, though that is rarely the case. (The truth is that a viable defence against cyberattacks can cost hundreds of thousands of dollars.) Effective cybersecurity requires a significant investment beyond hiring someone to manage it.

That's because cyberattacks affect a company on all levels—they require an immediate response, a careful consideration of the effect on corporate reputation and an adjustment of future growth predictions. When an acute crisis hits, the company first has to focus on the immediate practical problems, like endless help desk calls, a backlog of customer requests, and the possibility of being sued. Crisis management is the order of the day. At the same time, investors and owners are wondering how this hit to the company's reputation will hinder the growth of the company. Inside the company, managers and employees worry about their own jobs and liability; outside the company, regulators and society are probably already reacting. The dollar cost of the attack itself is probably one of the last things on people's minds.

Hiring a manager is only the first step. The manager may need to call on existing staff to take on new tasks; for instance, a network engineer might need to develop skills in network monitoring. Internal staff will be needed to build and execute solutions, and it may be necessary to hire external companies as consultants and to test the systems. If additional hardware and software are required, internal IT resources may have to be reallocated. These represent significant costs.

Nevertheless, many companies refuse to allocate sufficient budget funds to cybersecurity—either to fix problems caused by a known cyberattack or to prevent an attack in the first place. So CSMs often find themselves in the difficult position of doing what they can with limited resources, even if it won't be enough. If CSMs don't have the interpersonal skills and initiative to gain access to key decision-makers in the company to lobby for more internal and external resources, they simply won't be successful.

ORGANISATIONAL STRUCTURE: PUTTING IT ON PAPER

When considering a CSM applicant, try this:

Hand the applicant some paper and a pen and ask him or her to draw a rough diagram of the company's current organisation. Invite the candidate to explain how they would manage the structure. There's no right or wrong answer, but you will learn a lot about the candidate's organisational skills by listening to his or her response. If your request is met by silence, that is not a good sign—you may have a pure techie on your hands. The best candidates will have something to say about every piece of the chart.

What does a lack of success in this role look like? If the CSM is ineffective, the company's cybersecurity won't improve, and may even be diminished. That exposes the company's employees, customers, and shareholders to serious risks of breach, theft, blackmail, ransom payments, legal action, and more. What's worse, hiring a CSM without allocating sufficient budget funds can lead the company to have a false sense of security.

The consequences of failure are also considerable for the CSM. Most cybersecurity managers don't feel accepted, or even respected, by the companies they work for. If a cyberattack happens, the CSM is the one who gets blamed. If the CSM is terminated, she finds herself out of a job, and the company must spend time and resources to recruit and hire a new CSM to come in and fix the existing problems.

WHAT CAN A CSM DO TO SUCCEED?

Cybersecurity managers first need to understand what they will be up against. They need to go into a new job fully understanding the challenges we've talked about in this chapter, including lack of access to decision-makers, limited or no budget, unclear expectations, a nonexistent cybersecurity plan and a general lack of respect for their role within the company.

Next, smart CSMs will study the organisation to learn who operates the levers of power within the company. Who has the authority to draft a budget, and who can sign off on it? If budget decisions are made by committee, who is on that committee? The CSM can map it out on a whiteboard using an organisational chart or the company's website, including the different teams, leaders and stakeholders both inside and outside the company. The CSM should find ways to meet and interact with these key people, such as volunteering to participate on management committees and advisory teams where they can meet decision-makers. If security is even remotely relevant to the committee or team, management will invite the CSM to participate; directors and managers are becoming more aware that they need to involve the CSM. If the CSM finds they're not welcome in some groups, it's a problem; the CSM must have their support. They may need to work on the groups' attitude to open that door.

After this, the CSM should schedule meetings with key stakeholders to talk about the cyber risks involved in different areas of the company. The CSM's goal should be to help the managers and department heads understand the cybersecurity risks to their department and then convince those leaders to request the budget funds required to address the risks. Turning key decision-makers into champions for cybersecurity is one of the most effective strategies for acquiring the resources needed.

THE CYBERSECURITY PLAN

After meeting with each department head, the CSM should follow up with a memo to that person summarising what was discussed, the current risks and a prescription for fixing the problems. After meeting with all department heads and stakeholders, the CSM should create a full cybersecurity plan. In that document, the CSM can summarise his recommendations for each department in the company, assess the broader risks for the company as a whole, and include recommendations and a budget request. This will become a working document for the CSM. It should incorporate feedback and input from the stakeholders.

If the company denies the CSM's request for budget funds and later suffers a cyberattack, the CSM can point to the document in which he identified the weakness and proposed a solution. If the CSM wrote a memo requesting budget funds for system improvements and security testing but the request was denied, the attack can't be blamed on him. He didn't design the system.

However, the CSM can be blamed for *not* notifying the company of the risks he knows about. If he doesn't warn the company, if he doesn't communicate the risk, then any breach could be seen as his fault. The CSM can determine the best way to present the memo, either one-on-one, in a staff meeting, by email or in whatever way is appropriate.

The cybersecurity plan should be aligned with the company's overall strategy and goals. The language in the plan should start with those corporate business goals and then go into detail about how cybersecurity supports the organisational objectives. Do not include personal opinions in the document. This plan is a big-picture document that will resonate with the senior-most leaders of the organisation, including the CEO.

IDENTIFY THE INFLUENCERS

When mapping out the key stakeholders in a company, the CSM should not only pay attention to the people with senior titles, like CTO, CEO and VP. Sometimes people in other positions can wield enormous power. These people are influencers. For example, there might be someone in human resources who has been with the company for twenty years and has the ear of the CEO. For some reason, she has a lot of power. The CSM should be sure to identify and add influencers to the list of stakeholders. They will play a role in getting the CSM's cybersecurity plan approved.

SELLING THE CYBERSECURITY PLAN

Part of being a CSM means using internal sales skills. First, the CSM must sell the decision-makers on the idea of cybersecurity. This involves communicating the risks and the potential repercussions of not acting on those risks. What is the risk? How big is it? How can it be fixed? How much will it cost? With a detailed assessment, it's usually pretty easy for the company to make a decision.

The reaction to a cybersecurity plan and request for budget funds could be a resounding yes, a flat no, or anything in between. This is because different executives have widely differing risk tolerances. What seems like too much risk for one person is just another day at the office for someone else. Entrepreneurs tend to have a high risk tolerance, while bankers tend to want a very low level of risk.

CSMs should be aware that some people will view them as a threat and may resist them at every turn. They may have a known cyber weakness in their department that they've tried to fix on their own. Or perhaps they don't think cybersecurity impacts their department at all, so they consider the whole thing a nuisance and a waste of time. Whatever the reason, not everyone in the company will support the CSM or even welcome him or her. Tread carefully.

Regardless of the different reactions from stakeholders, it is the CSM's responsibility to own the cybersecurity plan and drive its internal adoption. If the CSM sits quietly in her office working on the computer, she'll make little progress. Instead, the CSM must proactively advocate for cybersecurity, even in the face of rejection.

> The CSM must proactively advocate for cybersecurity, even in the face of rejection.

Ultimately the CSM will need to get the plan accepted and implemented by the company. This decision may be made by the CEO, the board or a lower-level committee. The head of each department that will be directly impacted will also need to accept the plan and get on board.

Gaining acceptance for the cybersecurity plan often requires negotiation. There will be some give and take, and the CSM will have to revise and rewrite some or all of the document to please all stakeholders. Some stakeholders have more influence than others. The CSM should always listen to feedback from the CEO, for example, as their suggestions are usually insightful. CEOs know exactly where the company is going financially and strategically, and they understand the highest level of risks, such as strategic business risks. The CSM should incorporate the CEO's suggestions into the plan. Eventually, by being persistent and reasonable, the CSM will develop a plan that will be widely accepted.

EXPECTATIONS AND NEEDS VARY COMPANY TO COMPANY

The CSM needs to understand that organisations vary widely in their expectations and needs when it comes to cybersecurity. In general, bigger companies won't accept as much risk as smaller companies because they have more to lose. If a company with billions of dollars in revenue gets hacked, the loss could be far greater than it would be for a smaller company. Smaller companies are nimbler and can change faster, so they can react to cyber risks more quickly and effectively. This is why many startups hesitate to spend a lot of money on cybersecurity.

HIGH-LEVEL VERSUS LOW-LEVEL CYBER RISKS

Cybersecurity exists to protect and advance the goals of the business; that mission cannot be lost in the details. Unfortunately, it often is.

The quickest way for a cybersecurity manager to fall from grace within a company is to focus on smaller operational risks and low-level problems. Too much of that and leaders (including CEOs, VPs and other stakeholders) will naturally start to ignore the CSM because low-level problems aren't what's on their minds. They care about high-level risks that affect the overall business strategy.

Many CSMs have been relegated to the realm of 'IT geek' because they didn't focus on high-level problems. Once there, they tend to tackle every problem as an IT problem and neglect the people side.

The exception to this is a high-tech startup or growth company where the entire business is based on trade secrets that could be stolen, such as a startup built around a biotech process or medical technology, or a computer company with a proprietary algorithm that could be hacked. If those types of trade secrets are stolen, it could literally destroy the company. In those cases, even startups will take cybersecurity very seriously and spend money on it accordingly. If they can't afford it, they will get more money from their investors. Investors want to protect their investment in the company, so it's an easy sell.

Banks and financial institutions, on the other hand, always make security a top priority because they're holding clients' money. The challenge with these companies is that many of them are using old mainframe computers from twenty-five years ago. Legacy systems like that are extremely difficult and complex to upgrade to modern cybersecurity practices, and complexity is the enemy of security. In contrast, newer companies will operate on modern cloud computing systems, so updating their cybersecurity is much easier.

While we can identify general characteristics of some companies, like startups and banks, the truth is no two companies are exactly alike. Even an experienced cybersecurity manager who moves from one company to the next will have to start over and learn about the new company. The same tactics they used at their last company may not work in the new environment. The decision-making processes in each company are always different.

The stakeholders are different. The corporate cultures are different. The tolerance for cyber risk will vary by the size and age of the company, their current technology, their history of cyberattacks, the industry they're in and what part of the world they're located in geographically.

Even different departments within the same company have differing needs and expectations. Separate departments often run completely independent computer systems. Some departments may have been hacked, while others never have, so they have unique expectations, needs and perspectives.

COMPLEXITY IS THE ENEMY OF SECURITY

Most companies, as they grow, build unwieldy IT systems that make security extremely difficult. The CSM can have an impact simply by reducing that complexity. It's often a good idea to look for standardised solutions others are using and apply those to your company's problems.

Say your company has two hundred servers, each with its own protocols. The system is a nightmare to maintain. If you can show management how other companies have reduced similar systems to just two types of servers, your life will get much, much easier.

The company will be happy, too, because the existing complexity is probably a performance issue for them. Standardisation saves money.

IDENTIFY THE BIGGEST RISK FIRST

The best way for a CSM to understand and approach an organisation is first to identify the biggest cybersecurity risk. What are the company's leaders *most* scared of? What are they trying to prevent? Where is the company most vulnerable?

As the CSM investigates the various security risks in an organisation, they should keep a running list of the security problems they would ideally like to solve. If there are twenty items on the list, that's too many. With the list in hand, the CSM should rank items in order of urgency and importance, then begin eliminating the problems at the top of the list, moving on to lower-ranking items over time. That way, they can solve the company's most pressing security issues while making sure they don't spend too much time and too many resources on things that are irrelevant or insignificant.

PERCEIVED RISKS VARY

To prioritise risk, CSMs need to be aware of how the people in their company perceive risk and understand that the perception may not be driven solely by logic.

The biggest perceived risks may vary widely between two companies in the exact same industry. Some of the variance comes from leadership style—one company may be led by a risk-taker and another by someone less bold. Some variation arises from corporate culture. Two airlines with vastly different corporate cultures, for instance, may identify completely different needs and risks in cybersecurity.

The effective CSM must recognise a natural human foible, as well: people are notoriously bad at assessing risks accurately. For example, people worry more about living near a nuclear power plant than about crossing the street, even though they're much more likely to get hurt walking through town. A CEO might read articles every day about companies hit by data breaches, but if it hasn't happened to him in twenty years, he may think his company is immune, though it's not.

When preparing the cybersecurity plan, the CSM should focus recommendations on the highest-level strategic risks— the ones that will get the attention of top management. The CSM's goal is to be taken seriously, to have the budget approved and to get the cybersecurity plan implemented across the company.

CHAPTER TWO

KNOW THE RISK AND MOTIVATION FOR RESISTANCE

Risk is inherent in every decision that a person makes in life. When you cross the street, you might be run over by a bus. Eat at a restaurant, you might get food poisoning. Risk is everywhere, especially in business. In this chapter, we'll look at different methods for understanding and communicating those risks, especially in the face of resistance.

UNDERSTAND THE RISKS

Every company has different top risks, as does every department within each company. Banks want to protect their clients' money. Technology companies must guard their trade secrets. Hospitals must secure patient medical records. Sometimes there are risks that nobody has thought about yet. Sometimes a few people have thought about it, but leadership hasn't taken an active interest. The first thing the CSM needs to do is to discover the biggest risks the company faces. The best way to do this is by interviewing people inside the company. These conversations can uncover risks leadership may not be focused on but should be.

> Find your biggest single points of failure in your company and make sure management knows about them.

One fairly large company we worked with, for instance, had a few thousand staff and a few hundred million dollars in turnover. They had a huge enterprise resource planning (ERP) system that was integrated with their production—nothing worked without it. The ERP was mission critical, but there was no replication, no redundancy, and no backup system. If the system broke, physically or otherwise, they were shut down until they could fix it and make it work again.

Several people at the company knew about this, but management didn't register it as a major risk. They didn't do anything about it for many years. Why? Maybe they weren't so interested in the company's operations. Maybe they were more interested in their own sales. Maybe they didn't worry about it because they thought the ERP had been there for twenty years and worked just fine, so why would it break down now? Maybe they just had higher priorities. If you asked management in that company what the biggest risks to the continuity of their business were, they probably wouldn't tell you that the ERP was one of them. All the while, the company's existence depended on the efficient functioning of the ERP system. These types of risks—risks that could cause the company to go bankrupt—need to be identified first.

Lesser risks might be those that cause the company to lose efficiency or spend extra money or effort to operate certain systems or machinery. For example, IT is often perceived by management as a way to do things more efficiently than the competition. When the computers go down and the company loses the benefits of IT, their margins erode and their advantage over the competition is reduced, but they don't go out of business right away.

This is the language the CSM needs to use when talking about IT security and systems downtime. The CSM shouldn't suggest that the IT systems will never work again. They need to focus on the margin issue: the company will get slower at serving its customers or might not be able to serve them at all until the system is back up and running. Learn to speak the business language. Explain your security issues in terms of money, resources and time so management will understand it's not just about security—it's about running a company profitably.

FOCUS ON THE MOST RELEVANT RISKS

If the cybersecurity manager wants to be influential at their company and relevant in their work, they need to focus on the risks that are most relevant to that business. The CEO understands the specific risks the company faces. Whatever the CEO considers important is equally important to and must be understood by the CSM. That doesn't mean the CEO is always right; the CSM may need to educate the CEO (or CTO or CIO), but achieving alignment is essential because the CSM needs the high-level support. We can think of the CEO as an internal customer—the customer is not always correct, but they must be understood. CSMs should listen a lot at first, ask many questions, and learn how leadership perceives the situation. If the CSM achieves alignment with the CEO and acts accordingly, the CSM will be successful.

If there's no communication, that alignment is unlikely. In one case we witnessed, a security manager tried to buy a physical security access control and burglar alarm intrusion-and-detection system for the company's offices. The first offer the CSM got was four times more expensive than the cheapest system. The CSM, who was focused on the risk of a physical break-in, decided it was worthwhile to buy the more expensive system. The CEO, on the other hand, considered this purchase a way to be in compliance as inexpensively as possible. He preferred the cheaper system. In this instance, the cheaper one probably met the company's expectations and requirements better. In the end, the CEO decided to go with the cheaper option.

In this case, the CSM did not adequately understand the goals and objectives of the CEO. The CEO's objective was compliance, while the CSM's goal was optimal security. They're different. CSMs can take a simple lesson from this: they must communicate with company leadership and discuss risks internally. The CSM should always investigate the company leadership's priorities.

FALLOUT FROM RISKS

When CSMs begin to identify risks, they should think about what category of risk they're dealing with. Is the company worried about losing the secret sauce, the data, the trade secrets? Or is their main worry an interruption in the manufacturing line? If the process stops, how big a problem is that? Is it a financial risk? Is it a legal risk?

Sometimes the staff or teams will not be able to articulate their biggest risks. Discussing examples with them can help them identify risks that apply to their company or their department.

Here are examples of risks and how they materialise:

MANUFACTURING INTERRUPTIONS

Any company that manufactures goods knows the risk of the line shutting down. When the line stops, that costs money. There are myriad examples of production interruptions that can cause complicated problems up and down the supply chain and lead to massive losses.

Or if a company manufactures industrial paints, what happens when they need to print shipping labels for all the places the paint is going—and the printers fail? If the printer doesn't spit out the labels, nobody will know which lorry to load the paint onto. They could wind up with empty lorries waiting outside and a full warehouse of paint inside since production doesn't just shut off immediately. They need some kind of redundancy or backup plan.

Manufacturing interruptions come in all shapes and sizes. But almost all of them are time-critical.

LOGISTICAL DISRUPTIONS

Logistics is a risky business on a good day. Companies deal with constant deadlines, uncontrollable weather conditions, mechanical malfunctions and a thousand other risks every day.

Let's say a logistics company has to deliver parcels for a client. The package someone sends to their friend goes through a huge automated facility where the parcels are sorted and put onto lorries, with lorries always coming in and out.

Now let's say a design error in the network and a DDoS (distributed denial of service) attack hits the network. It prevents not only external data communications but internal channels as well because the same network was handling both internal and external traffic. The system is at capacity and cannot handle any more data.

That plant can't run. Nobody can deliver parcels. But lorries will keep arriving. They keep offloading parcels into the facility. Meanwhile, automated sorting lines won't run, and machines won't sort. The lorries are full but can't leave. Within hours, the facility is overflowing. Now imagine if that one facility handles about 70 percent of all parcels in that small country. It could be a major disruption to the economy.

This nightmare scenario is not unusual in logistics. The risks in logistics are high, and backup plans are critical.

DATA BREACHES

If companies fail to prioritise business continuity risks

like the above, they're at least as likely to overlook 'external' risks like the potential loss of customer data. We once worked for a company that should have known better; they provided healthcare services for private individuals. The company was developing better IT services for its customers—solutions that would monitor their operations, compliance and security systems and notify them of any event that looked like a breach. Great idea but for one problem: they stored all the information in a centralised log system and left it wide open online.

BREAK THE CYCLE

Companies can get used to having business interruptions. Operations break down, the company finds someone to blame, they fix the problem, and they wait for the next time. It's time to break that cycle. Find the most severe problems and fix those first. Even if you can't fix them, you'll at least see the next crisis coming.

Someone in the public stumbled onto the problem and notified the company. It turned out the developers had left the communication ports in the firewall open. All anyone had to do was connect and download, no username or password necessary.

Too many companies fail to understand what they're doing when they use virtual servers. 'The cloud' is a mystery to them and one they may not investigate until it's too late. In this case, the situation lasted over a year. Because the company had no way to know if something was or wasn't stolen, they had to live through the nightmare of notifying every single customer.

COMPLIANCE CHAOS

Compliance, even when intended to prevent problems, can actually create more disruptions than it solves. In reality, companies rarely benefit from compliance outside of being able to market themselves as compliance-certified; it seldom helps secure the company. Worse, it doesn't even have to make sense. For example, we once worked with a company that provided IT services for many companies all over the world. They wanted to make sure their vendors followed proper compliance, so they conducted frequent facility audits. They would send auditors to inspect the data centres and make sure they were compliant.

A problem arose when compliance declared that there couldn't be any external people (i.e., the auditors) inside the data centres—only internal people were permitted to be there. Ironically, the company would fail to pass compliance because there were too many auditors auditing their facilities. This is a strange but telling example of compliance risks in IT.

APPLY STANDARDS SITUATIONALLY

Compliance standards require CSMs to do certain tasks and take certain cybersecurity measures, but not all of them are necessary all the time. If the CSM can demonstrate that a requirement is not applicable to the organisation because there's no risk, they can save time and money because they won't need to implement it.

Consider the example of the Russian company that handled credit card information without addressing requirements specified by the American credit card company that hired them. They looked at the list of requirements, threw up their hands, and did nothing. The American company was incensed; they threatened to cancel the deal.

Should the Russian company have checked every box on that list? Maybe not. It's possible that they could have avoided that by analysing the requirements to see which were truly applicable and which were not. By documenting the reasons some steps were inapplicable, they might have been able to reduce the number of tasks on their list. Then, in communication with the American company, they might have been able to complete only those items.

Compliance is mandatory, but sometimes it's malleable too.

Whether requirements come from laws, contractual requirements with another company or internal requirements, companies should try to follow them, especially in IT. The bottom line: you can't ignore compliance.

FINANCIAL RECORDS BREACHES

The banking and finance industries are based on trust.

When a financial institution has a breach, it goes to the core of the business—they lose customers' trust. Let's say a wealth management company or small bank loses their client records. The people who came to them to ensure their wealth was safe, secret and exclusive may now leave. New customers may be reluctant to come on board. That's why the banking industry is the first to implement security measures.

This happened with the Panama Papers scandal. You may recall this hacking incident from news reports: millions of documents containing detailed personal financial information on wealthy individuals and government officials leaked to news outlets and the public. Some documents included information on banks that were illegally helping their powerful clients hide money and avoid taxation. After that data leak became a public scandal, potential customers shied away because of the leaks. Not only did wealthy people avoid the organisations involved and likely pulled their money, but average people avoided them, too, because of the perception that they were participating in shady tax planning.

EXTERNAL PROVIDER DEPENDENCE

Human resources maintains employee records with very personal information: salary, social security numbers, banking information and so on. Because the employer has to store this information for employees, they are responsible for storing it securely.

At the same time, payroll and other management tools used in HR are often outsourced to external service providers. The risk is clear—if there's a leak or data breach at the provider, the company could lose their employees' personal information.

This would be a disaster. First, the company would face fines. They may also be required to acquire credit monitoring services for every individual whose personal information was subject to a leak. Companies can also get sued for enormous sums by their employees, with the amount varying by country. (The United States and Germany have exceptionally high fines.) The financial impact can be huge, not to mention the impact on the company's own employees in terms of reputation and esteem.

BUILDING TRUST

In today's world, most companies are part of an ecosystem of companies that provide resources to each other and drive responses from each other. Acknowledging this interdependency is powerful. Use it to build trust.

RETAIL RISKS

Credit card fraud is common in retail. Companies use payment systems from specialised vendors which may or may not be secure. They might acquire a payment system for cashiers, for instance, which includes a point of sale (POS) system so the customer can pay with a credit card.

Criminals are breaching these types of networks by phishing or other means. Once they gain access to the network, they create specialised malware for these POS devices. The malware collects credit card details and transmits them out to the internet for criminals to save and store for later use. Then the criminals sell that data to third parties, who use it for illegal, online, card-not-present purchases.

Then the shop and the bank that issued the card argue over who's to blame and who should pay. If there is $50 million in total annual credit card fraud at the bank, who should pay? The bank or the processing company? Or the careless customer who didn't take proper precautions? The costs can be huge, even before the company begins to address the cost of solving the breach and replacing all the stolen cards. The feuding parties will likely try to come to some kind of independent agreement because the last thing a company wants to do to solve their business issues is air them in court. That's bad PR for everyone.

INTELLECTUAL PROPERTY THEFT

A startup that's building an entire business on a single proprietary technology has the potential for high reward but also has high risk. The biggest risk is intellectual property theft. If the startup loses its secret sauce, whether it's hardware, software, medical, biotech or some other trade secret, that could destroy the competitive advantage and ruin the company. For example, consider a silicon-chip manufacturing company with a proprietary process for producing precision chips used in sensors for cell phones, smartwatches and even satellites. They succeed because they own the intellectual property behind that innovative technology. If someone steals it, they steal the innovation and deal a death blow to the business.

When a startup relies on a single technology, the risk is incredibly high. If they lose it, they lose everything. Startups usually don't have much support available. They don't have a massive database of clients yet or a complex network of relationships or partners they can rely on. The value of most startups is the technology. It must be protected at all costs.

RESISTANCE TO CHANGE

There are so many risk scenarios that can be mitigated by well-executed cybersecurity management, it might seem like an easy sell. Still, it would be unwise for a CSM to boldly walk in to a staff meeting one day and announce, 'We are revamping all the security procedures in this department, implementing a bunch of new security processes, and changing the way you've been working for the past ten years.'

No matter how strong the case for change, people will resist, either because they fear new things or they have come to rely on the consistency of their old ways. Or both. Not everyone admits it; some people outright deny that they fear change or try hard to downplay their discomfort. Others get angry. When a CSM encounters resistance to a new policy or procedure, they should try to find the psychological root cause. If the CSM understands what drives a person, they will be better able to address the underlying issues. Here's how:

USE THE RIGHT REWARD

One way to determine what motivates people is to notice

how they like to be rewarded. Different employees respond to different rewards for a job well done. Some appreciate being given more freedom and fewer rules. Others value getting more time off to spend with their family or to work on pet projects. Some people seek safety; they're motivated by reliable forecasts about what's going to happen in the future. Some want to get information before others. Many people are motivated by money, while others are inspired by awards and recognition.

CHANGE IS GOOD

If you can explain to the person why a change is good for them personally, they usually accept it. It's even better if they figure it out themselves.

Sometimes, all a CSM has to do is ask, 'Can you tell me what you do in security in your job?'

After they answer, most people are ready to get on board with the proposed changes.

The attention is a type of reward.

If people are motivated by money, the CSM might emphasise that a more secure company means more secure jobs. Others might be motivated by safety and forecasting, so the CSM should explain how the new security measures ensure there won't be any surprises. For people who are motivated by having more freedom, the CSM can show how the new security measures will give them fewer hoops to jump through. In all cases, the CSM must employ empathy to understand how people feel, not to manipulate them but to serve them better.

COMMUNICATE WHY

Too many security managers think of their job as just writing up policy and procedures: 'Don't click on links in emails from unknown senders. Change your password at least every thirty days.' Then they put those policies on the internal network and hope that people read it. And that's about it.

That's not enough. You have to explain the why as well as the how.

If a company just wants to meet the compliance requirements and provide some training for employees, a list of procedures might be enough to protect them from getting fined, but it won't actually make the company safer or change people's behaviour. People seldom act differently unless they know *why* they are being asked to make the change. The CSM must help employees understand that they have shared goals. They all want to make sure that risk is minimised; the CSM wants to secure the organisation, and so do the employees. The policies and procedures are intended to help everyone meet that goal.

One way to demonstrate this is to give a detailed example, like hackers who get in through malicious emails and phishing. The CSM can explain to employees how these attacks work by offering an example. Say the probability of a single user clicking on a phishing message link is fifty-fifty. Knowing that, the attacker is able to calculate how many messages need to be sent to have a 95 percent probability of success. So they might send twenty messages to different employees with high confidence that someone in the company will fall for it. One employee will click, and that's all it takes. That's why everyone has to be trained well. Examples like this can help employees understand that there is a reason for every security policy.

LET THE NUMBERS SPEAK

We recommend coming up with two numbers to help explain why security policies exist and must be followed. The first number is how much can be lost, and the second is the probability of losing it. If, before security training, 80 percent of employees click the link in a suspicious email, but *after* awareness training, only 20 percent click the link, we've reduced the problem quite significantly. This reduced risk actually creates a financial gain, in a way, by avoiding losses.

Using numbers, the CSM can demonstrate that, compared to the cost of a breach, training employees is actually quite cheap. It's probably well worth spending $10,000 for training if the risk is reduced by 60 percent. Using that sort of cost-benefit analysis, mapped out in risk language, can help everyone understand that security training is a wise investment.

Every cyber risk can and should be translated into risk language and numbers. The numbers don't even have to be exact. Ballpark numbers work. Without understandable quantification, few people will take the threat seriously, even fewer decisions will get made and little change will occur.

COPING WITH RESISTERS

CSMs in large companies generally interact with three different types of people: those who help them and get on board right away, those who go along only if others do and those who resist. The resisters are the most difficult to deal with. If a CSM doesn't do anything about the resisters, they're in for a tough battle, and they might even fail. Ignoring naysayers is the worst thing to do. Instead, CSMs should identify those people and turn them around so they won't interfere and might eventually come to help.

The first step with resisters is to identify and acknowledge them. This step alone may resolve tensions. If the resisters still don't change their behaviour, though, the

CSM might need to go to a superior to establish authority. If the boss communicates the need to staff and lets them know what's coming, it creates a sense of inevitability around the change. If the person knows the change is going to happen anyway, they will realise their only choice is whether to help or not.

> Nobody wants to be the last resister.

Parents understand the value in teaching kids that some things are just inevitable. Instead of asking, 'Are you ready to go now?' they'll say, 'Okay, we're leaving for Grandma's house in five minutes. If you don't get dressed right away, you'll be riding in the car in your jammies. We are going to Grandma's house.' The kids usually get dressed. The interaction doesn't have to be uncomfortable or threatening, just a statement of inevitability. If it works for kids, it usually works for everybody else.

We recognise that it's never an ideal situation if the CSM has to go to the boss, but sometimes it's the only way. If the change the CSM is proposing is important for the company and will address significant risks, it should be a joint effort that everyone gets behind. At the very least, no one should be interfering. The bottom line is, someone who is making security more difficult poses a threat to the company.

Motivating people who resist that change is a necessary part of the job. The more experience a CSM gets, the better they will be able to navigate change. In the end, the work the CSM does is for the good of the company. Eventually everyone will realise that and get on board.

CHAPTER THREE

KNOW THE STRUCTURE AND KEY STAKEHOLDERS

A good CSM must understand both the organisational structure and the power structure of the organisation they work in. This is essential. If the CSM doesn't know who makes the final decisions, who controls the money, and who the key influencers are, it will be difficult to be effective in the CSM role.

STUDY THE ORG CHART

Studying the organisational chart of the company is the quickest way to get the lay of the land, figure out who's in charge, and discover who reports to whom.

Looking at the org chart, the CSM can see if the company is structured in a line or a matrix. In a line organisation, depicted by the typical upside-down tree chart, the hierarchy is clear. Key executives are listed at the top of the chart and department heads below them, then the managers, and so on. In this scheme, each owner has a budget, and you can go straight to the owner with requests. In a line organisation, the CSM should talk with each owner about how security is relevant to them and their department.

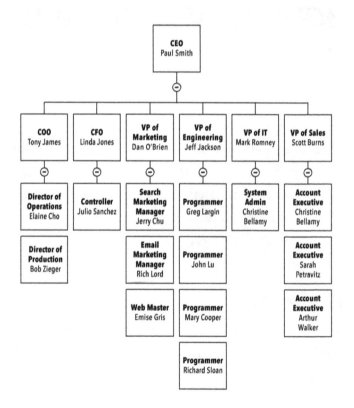

In a matrix organisation, however, the landscape is a bit different. Here, employees are divided into teams by projects, which gives workers multiple reporting relationships, for example, to a functional manager and a project manager. In this scheme, a quality manager or product owner might work with people on every line, making for some complex relationships. What the CSM needs to understand is that this manager or owner is not the boss with the budget. In a matrix, the CSM must negotiate with people in all of the business lines.

In either type of organisation, the CSM's goal is to find out about any security concerns and learn who is responsible for those areas. It's not always clear. Many supervisors, though they are responsible for monitoring safety and security, don't even know where to find cybersecurity procedures. They may not know what to check when they hire people, or they might forget to collect computers when someone is fired. Some don't feel security is their business—production managers figure they handle production plans, not security—until the day there is a problem and they realise it is their responsibility.

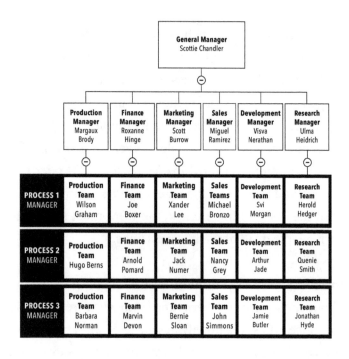

	Production	Finance	Marketing	Sales	Development	Research
	General Manager Scottie Chandler					
	Production Manager Margaux Brody	**Finance Manager** Roxanne Hinge	**Marketing Manager** Scott Burrow	**Sales Manager** Miguel Ramirez	**Development Manager** Visva Nerathan	**Research Manager** Ulma Heidrich
PROCESS 1 MANAGER	**Production Team** Wilson Graham	**Finance Team** Joe Boxer	**Marketing Team** Xander Lee	**Sales Teams** Michael Bronzo	**Development Team** Svi Morgan	**Research Team** Herold Hedger
PROCESS 2 MANAGER	**Production Team** Hugo Berns	**Finance Team** Arnold Pomard	**Marketing Team** Jack Numer	**Sales Team** Nancy Grey	**Development Team** Arthur Jade	**Research Team** Quenie Smith
PROCESS 3 MANAGER	**Production Team** Barbara Norman	**Finance Team** Marvin Devon	**Marketing Team** Bernie Sloan	**Sales Team** John Simmons	**Development Team** Jamie Butler	**Research Team** Jonathan Hyde

For example, we worked with a government organisation in Finland where an older lady, who had served as the cybersecurity manager, was preparing to retire. As she prepared to hand over the reins, it became clear she didn't have a clue about the organisational structure or even where to report security issues.

After a day or two working with this lady, studying the organisation, and having discussions with her and her colleagues, it was actually quite clear who was making decisions in the company and what kind of management structure was needed for security. This woman could have studied the org chart and discovered who she needed to talk to. She had just never bothered to do it.

Finding out about the decision-making process in an organisation is crucial. Without a solid grasp of organisational structure and decision-making, the CSM will struggle to get anything accomplished. The organisational chart is a helpful starting point, but it's equally important to study process flow charts and understand how things are produced and communicated.

With that in mind, CSMs also need to be aware that all the information about who has power and influence in any organisation is not necessarily on the official org chart. Often, in bigger companies especially, virtual teams have authority to decide on certain matters, especially regarding security. Those teams aren't usually visible in the chart. Many influencers may not appear on the chart at all, so CSMs have to talk to the people in charge, ask them how decisions are made and find out who makes the final decision.

THINK BIG

The larger the company, the more challenging it is to communicate security issues and to get acceptance on policy. In a smaller company, CSMs can always go right to the head of the company to get a yes or no. That's easy.

But when working with a bigger, more complex organisation, it can be harder to discern the existing roles and power structures.

For example, firewall management isn't just about managing a device; it's about managing people. The firewall is centralised and everyone goes through it. Employees working from home, for instance, have to be able to access internal applications held by different business units, and they need a way to do it securely. Firewall rules need to allow access while remaining secure.

There may also be policies that get created in different, cross-functional bodies. Geography may be a factor—say a Singapore company's IT or management is in another country and needs permission to do something in Singapore. Seeking permission causes delays the organisation can't afford because the problem is real and happening now. Hackers are not going to wait for permission from the US or Hong Kong.

Multinational companies have to figure out how to address this problem. Are they going to write policy so it applies to all countries similarly? Do they want to leave it on a high level so it's floating in the ivory tower of the home office? Or will they start making a different policy for each country? Every decision, whether centralising security decision-making power in the main office, or spreading it out to each country's local organisations, or working out a hybrid model, will have significant ramifications.

Understanding the differences between countries and organisations and their managers is crucial for CSMs because they need to know who holds decision-making power. For example, a company in the US with offices in Bangkok might not be very well managed in Thailand. Let's say the Bangkok office is missing a CEO at the moment, or maybe it has financial problems or sales problems. In that case, it would be almost impossible to push for the security issues from HQ because it's not generally managed well there.

IDENTIFY THE INFLUENCERS

With so many people's input to consider, the CSM has to make some decisions about whom to listen to first. The dog that barks the loudest is not usually the most dangerous one. In this case, the analogy means that some of the most powerful people in any organisation may not have any official role in the management structure. These are usually long-time employees who get things done and influence decisions, even though their title might be administrative assistant or office manager. Most companies have these influencers, people who are superconnected to the highest-level executives and know them all well. These influencers are important people to identify.

OVERLOOKED INFLUENCERS

Sooner or later, someone in your company will ask, 'Do you know Martha? Martha knows everyone!' If you don't know your company's 'Martha' yet, get to know her. Martha might be an executive assistant, a secretary, or a human resources manager. No matter her official position, she knows what's going on and can get your ideas heard.

It might be difficult to identify these roles before spending considerable time in the office. Getting to know people is crucial, and that takes time. Sometimes, a CSM will get lucky and get to work with a very influential person who supports cybersecurity and supports the CSM's efforts. For example, we know a multinational company in Europe that had a fairly large IT budget—in the tens of millions. It just so happened that a lot of key decisions were made in the application development steering committee. The steering committee didn't appear anywhere on the org chart, but that is where the decisions were made.

When a CSM is building an expensive application for a business purpose and has the ear of the committee that makes decisions about those applications, they can control a fairly large budget. The CSM in that scenario had better be present in those committee meetings to be heard and to find out how the company is led and what the foundational issues are.

CSMs may be surprised to discover what influences a company's decisions. Even a past event can be an influencer to a company. This includes any major crises in the past—such as a past cyberattack, a failure in compliance, a bribery scandal, or any negative story in the media. After events like that, security will be more important than ever, even if the scandal wasn't directly related to security.

KNOW THE STAKEHOLDERS

Every organisation has multiple stakeholders, and the CSM would be wise to identify them and get to know them, because each type of stakeholder can pose a unique challenge. At some point, the CSM can expect to navigate some kind of fight between the manager and the compliance folks, for instance, or between the manager and the risk folks. Some challenges are predictable, and the CSM can prepare for and manage those. In this section, we'll review some of the most common stakeholder challenges.

ALIGNING WITH COMPLIANCE-DRIVEN STAKEHOLDERS

Compliance means every company in the same field has to solve the same requirements. These tasks come from industry standards, or even laws, and are imposed from outside the organisation. Companies have to address them even if they don't seem to make any sense. Fortunately, companies do have some choices about the solution and its implementation. For instance, if user access management is a requirement in a standard, you have to meet it, but you can choose how to do it, whether through policy, training, installing new technologies or other means.

It's easy to lose sight of this relative flexibility because people who are experts in compliance tend to take things very literally. If compliance says to open the window, you must open the window, even if there's already a massive hole in the wall or an open door one metre from the window. Nobody gets bonus points for going above and beyond in compliance. They just need to tick the box.

And yet compliance is unavoidable because if companies are non-compliant with regulations, they run a risk of eventually losing their licence to operate in their field of business. The process may take a while—companies usually get some time to respond and try to fix the issue. On the other hand, public response can be immediate. Airlines, for example, can't skip their safety check—even if they would save a ton of money—because they would get very bad press and go out of business.

Compliance is such a challenge for business owners because regulators don't care about value. Most compliance officers don't even talk about cost-benefit analysis or return on investments; they might not even know what that means. (Although a deeper look often reveals the return is in the regulator's interest.)

From a business perspective, compliance gives no competitive advantage because everyone is in the same boat. Compliance almost always destroys stakeholder value within the company. The business-minded CSM will take compliance with a grain of salt and try to implement it so they can stay in business while doing minimal damage to the bottom line.

The irony is that all the compliance in the world doesn't always lower your risks. Complying with all the regulatory standards doesn't make a company secure. You can buy a scanning tool for your e-commerce site and run it four times a year—you'll be compliant. You won't get fined. But you may not be any more secure than before. Still, if a CEO has to make a choice between compliance and security, which one should he make? There's really no debate—he has to always choose compliance.

ALIGNING WITH RISK-DRIVEN STAKEHOLDERS

Risk experts have a very different mindset from compliance officers. Risk experts believe that risks are inherent in everything, so it's their mission to list and assess them all. Even if there seems to be an unlimited amount of risk in a given industry, their job is to attempt to analyse it. They want to be helpful. They try to understand the business. Risk experts are often much easier to talk to than compliance officers.

Risk people are numbers people who calculate probabilities. Risk management has its roots in insurance, where most risks are quantifiable. Insurance companies look for the magic formula that makes the most money, and they spend years creating it. If they get the life insurance formula right, they make better profits than their competitors. They don't even start insuring something that they can't calculate. Unfortunately, finding a formula is seldom straightforward. Most of the risks in the world are hard to quantify.

The big problem for companies is that their risk expert is usually overwhelmed by the many people who come to them with their risks. They can only manage a few of the biggest ones, leaving the rest behind. Large organisations with thousands of employees might solve this problem by hiring multiple risk managers with a variety of titles, like senior risk manager or director of risk management. They might even benefit from creating a risk management team. In some industries, specialised roles may be dedicated to different areas; for instance, people with the title of risk manager may have specialised functions that deal with fraud or assessing the credibility of cardholders before giving them limits. In those roles, they're dealing with one particular type of risk.

While compliance people want you to tick the box, risk management people would rather see you run penetration testing or some kind of analysis to find out if the application is actually secure or not because they want to mitigate the risk. They also want to maximise opportunities. For example, building in strong security while you're creating a new internet trading platform does a lot more than just prevent losses—it gives you the opportunity to be in business. Without it, customers will not trust you. By implementing it, you not only prevent losses, you enable the whole business.

ALIGNING WITH BUSINESS-DRIVEN STAKEHOLDERS

Business people can push back on both the compliance and risk experts. They're interested in how fast things can move and how much profit they can make. Once they've developed a new system, they want to know: Is it safe to put online now or not? How long will it take to fix it? How do they avoid delays? Business is always about profit, and it always comes with a risk, so a CSM who understands both can help make the best decisions.

Business people are goal-oriented; they don't want to hear about problems. The last thing they want to hear is compliance saying, 'You didn't have this one policy written up and accepted by certain boards.' Business people are always trying to accomplish more, and it looks to them like compliance sets obstacles and limits just to rein them in.

BALANCING STAKEHOLDER INTERESTS

Each category of stakeholders comes with its own deadlines and pressures, and they often conflict. If there's a potential security problem, for example, the compliance people don't care how long it takes to address it, while the risk people try to figure out the probability that someone will hack it before a patch is in place. Business people, on the other hand, might think it's not a big deal at all. They're apt to say, 'If we put it online now, we're going to make this much profit, so it doesn't matter if there's a small loss.'

The CSM in any organisation will have to interact with a variety of stakeholders like this, each with their own set of wants and needs. The art of CSM is in trying to find the right balance between these three groups.

CHAPTER FOUR

KNOW THE CURRENT LIABILITIES

Before the cloud, everyone had his or her own little medieval IT castle—the data centre. Much like a real castle, it had a perimeter wall (network edge), a gate with guards who allowed or blocked access (firewall) and different defence zones within the walls (network segments). It was pretty simple to identify liabilities—they came from the guys trying to ram in the castle door.

Today, the perimeter is gone. We live in a network of small villages where the data now resides. An attack could come from anywhere. Most people, especially if they've been in the business for a long time, have not moved to the new paradigm yet. They're not thinking about outside services, but they should be. It's the CSM's responsibility to shift that paradigm so people are thinking about vulnerabilities from the attacker's position, not from a supposedly safe spot inside their own castle walls.

To do that, the CSM needs to know how much the company has been exposed to security risks, now and in the past. This is *cyber exposure*, and it doesn't just mean data leaked to the public or breached from the organisation. Data can be leaked from partners, the supply chain, employees, clients or anyone with access to company information.

TYPES OF LIABILITIES DUE TO CYBER EXPOSURE

We have met a number of CSMs who had long careers in their companies. They got very nervous after learning that they never knew about many of their liabilities and cyber exposure. Now they wondered, had they been hacked three years ago but the CSM was never informed?

If you're new to the job, make sure you won't be surprised about the past. Do your homework about exposure so you're not blindsided by people coming to you with exposures. As the security expert, the CSM must be the one to inform others. That's the job.

If a CSM has the means to learn about their liabilities, they should use that means. Surprisingly, many people decide not to do so; they decide not to be informed. That's sorely misguided. Here's some of what they are missing:

The liabilities can be split into eight different categories or sectors. Any new cybersecurity manager needs to know if any of these things has happened already—or, in the worst case, is happening on their very first day on the job.

DISCLOSURE OF SENSITIVE INFORMATION

The first category of liability is disclosure of sensitive information. Every company has internal emails, documents, data and pricing. Tech companies have source code, proprietary software they're developing and trade secrets. If this stuff gets leaked out, it is first and foremost a reputation risk. This must be looked at from two perspectives: what it means for the business and what it means to risk management people.

EXTERNAL DATA BREACH

The second category is an external data breach. This is a data breach through third parties—business partners, supply chain vendors, HR, payroll companies, payment processors, and financial institutions. It's a theft of company information, and the company doesn't have any control over it. For example, if HR gives their employees' private information to an outsourced payroll company and that system gets hacked, every employee's address and bank account numbers could be breached.

Once the data is stolen, there's nothing the company can do. The data is out there, and it's no longer within their control to protect it. The liability in these cases will be viewed as the company's, not the external service providers. In the eyes of the users, the data was in the hands of the company and it was stolen—they don't care that a third-party vendor was the source of the leak; they trusted the company to protect their information. They will say, 'I gave the information to you, and you lost it.' This category of breach is common with cloud service providers.

BLACK MARKET ACTIVITY

The third category is black market activity. This happens when people sell company information to the highest bidder, usually on the dark web. Black market activity might be initiated by a rogue employee who's offering to sell information for money or by an outsider soliciting an employee to sell them information. There are dark web forums that specialise in inside information trading. These marketplaces allow users to say what kind of information they want and ask how much it would cost. Then the hackers start bidding: 'You pay me $250 or $500 and I'll get all fifty thousand customers' home addresses and social security numbers for you.'

Black market activity is dangerous, even when the information hasn't been lost or breached. Just the fact that someone is saying publicly that they are willing to pay for or sell this kind of information makes it more likely such a transaction will take place in the future. An alert cybersecurity manager wants to know if their company is being talked about on the dark web. Whether the data has been lost or not, you need to be aware if someone is talking about you. You need to know if you're a target.

EXPOSED FINANCIAL INFORMATION

The fourth category is exposed financial information. If the company is sending money to someone else, there's a transaction log, and these get leaked quite often. Usually, transactional information is stolen, but thieves may also walk away with complete credit card or bank account numbers.

It sounds like it might not be a big deal—companies send millions of dollars to one supplier and thousands to another. But if hackers know how much money clients are routinely sending to which companies, they can pretend to be those companies and send out fake invoices or emails claiming the customer should have sent a larger amount, thus duping the company into paying a fake invoice to the hacker's bank account. Or if a company sends $10 million every month, one month they might get the same invoice with a different bank address. People are routinely selling information gleaned from credit card access and database dumps that are stolen from banks with all of their customer records and transactions at the ready. A CSM needs to know if this has happened already.

EXPOSED CREDENTIALS

The fifth category is exposed credentials. This includes passwords, usernames, combinations or some kind of security token. When hackers steal this information, they can access company systems. The challenge is, it doesn't matter how well you're secured, how many systems you have, or how many people you have working on your cybersecurity team. If someone gets the password, they can log in, and no one will be aware of it because it's a legitimate login.

Most employees will reuse their work-related usernames and passwords in external services. Usernames are typically based on email addresses that contain the domain name suffix of the company. It's a little like tagging your house keys with your home address. When your keys get lost, the burglar who finds them will know exactly where to go. When company credentials are stolen, it's the same story—the thief now knows how to access your company.

This is nearly impossible to control because it happens completely outside of the company. There's no way the organisation can watch what's happening everywhere. This exposure is beyond the company's reach.

PERSONAL INFORMATION

The sixth type of liability is personal information. This includes people's names, physical addresses, social security numbers, hospital data and even personal hobbies. The risk here is identity theft and fraud. If a thief has enough information about a person, they can open a bank account, get a payday loan or credit accounts to make fraudulent purchases.

Sometimes hackers get all the information they need from data breaches, and sometimes they have to search for more to complete the picture. If the information is leaking from the organisation, hackers can construct complete profiles of people and then use that to further hack into company systems or to buy things.

Identities are hard currency. They enable synthetic fraud. In this type of fraud, criminals use personal information to construct real-looking profiles and sign up for services, make unauthorised purchases and so on.

Also, consider this. if you get an email from your boss from his LinkedIn account, and you check that account looks like his, then agree to click a link, how can you know if you were fooled by a spear-phishing email? Or was it actually from the boss? Most people have no means of distinguishing a well-crafted fake profile.

HACKER GROUP TARGETING

The seventh category is called hacker group targeting. There are many hacker groups online, and they don't even know each other, at least not their true identities. They're all anonymous, hiding behind avatars and nicknames. Even though they want to remain anonymous, hacker groups still want to communicate with each other. They talk about what company they want to target next, and some groups even publish their targets months before an attack. If a company is listed as a target, they need to know before it happens.

Typically, hacker groups launch DDoS attacks against a company's critical e-services or its website, disrupting normal traffic. While individuals are most often targeted by spear-phishing campaigns, companies suffer more at the hands of DDoS.

INTERNAL DATA BREACH

The last category is the internal data breach. All the other categories we've discussed are from external sources, but this breach comes from internal systems. This is a breach perpetrated by someone inside your own organisation, system or database. For example, it could be your Windows domain controller machine was accessed by an employee who dumped all the accounts, plus all the documents, and published it on the dark web for a profit.

So, seven of these categories occur outside of the organisation and one inside the organisation. In a way, it really doesn't matter who's causing any of these categories of breaches because the damage is the same. The cybersecurity manager needs to find out about all of them when they start the job. That information will inform policy and decisions going forward.

TAKE CARE OF YOUR PEOPLE

Companies need to be concerned about their people even when they're operating outside the company's own systems. Attacks won't necessarily come from an employee's work email; they can also come from a Gmail account or a LinkedIn exchange.

Companies need to care about their people and even their families. These concerns are no longer limited to corporate territory. It's not just about work anymore; it's about the whole society.

Using this eight-fold model can be extremely helpful—with it, you will find out about your exposure from an outside attacker's perspective—but it also creates liability for you as a CSM. Now that you know where to look for problems, you're going to be held liable if you don't discover them.

MIGRATION OF DATA TO THE CLOUD

Ten years ago, an IT setup would have looked like this: internal company systems resided in servers on the company's physical premises and were protected by a firewall, database servers and some workstation networks, and these were divided into zones containing all of your secure information. That's an outdated model. There are very few internal mainframes still operating. Most companies go asset-light. Maybe some banks still do things that way, but most companies have migrated over to using cloud services.

Cloud computing has changed everything. By using the cloud, companies can store their information outside of the fortress, usually in tens of different cloud services. All the new business applications they build are based in the cloud. Companies often prefer now to buy software as a service (SaaS) instead of building their own solutions; it's cheaper and faster. Information is becoming much more externalised. As you can imagine, this creates challenges for the CSM. One challenge is that while data is put to these SaaS cloud systems, the old security controls that used to apply in the fortress model no longer apply. Say you want to scan vulnerabilities or do a penetration test of your application. Now that it is in the cloud, that probably won't be possible. Even if you can do it, with the permission of the vendor who runs the SaaS application, you probably won't get useful results. Instead, CSMs should look for solutions that focus on monitoring and protecting the data instead of system vulnerabilities.

It's important that the CSM understands what kinds of systems the company has, what kinds of information are in each system and what kinds of services they provide to those systems. This can be done with technical tools, and usually some in-person interviews are needed as well.

The latest regulations in the European Union and the United States require that every large company has a data protection officer, but the CSM can't rely on them. Data protection officer is a different role than CSM. The data protection officer is there to find out where confidential customer information is stored in the company and to protect it. They are responsible for controlling and securing it adequately, not doing the protection or design work that will help implement the security processes. The CSM needs to know just as much as the data protection officer and much more besides.

> What is the cloud? It's just somebody else's computer.

LOOKING IN THE RIGHT PLACES

Most security managers spend about 80 percent of their time securing the company's internal assets. They look very little at the outside, even though critical data for any company always lies outside the organisation. Seven out of the eight liability categories we listed above are external, so most of the CSM's efforts should look outside.

For example, fewer and fewer companies now own their own IT systems and networks; they lease them and use cloud systems, outside maintenance, and outside development. They pay for what they use and give the control to a third party. When they give the control out—for instance, letting a partner operate the service desk—they must monitor it carefully. Handing information to that third party, the company becomes responsible for monitoring that third party's security. If the information is stolen, the company is liable. If they didn't monitor or even vet the third party, they are even more liable.

Few people understand the extent of the liabilities and the exposure they take on when services are outsourced. This is one of the biggest challenges CSMs face right now. The misunderstanding leads to arguments over who is to blame for a breach. Is it the board's problem? The owner's problem? A management problem? This gets especially tricky if management remains silent for fear of revealing how much data has been exposed already.

The CSM could be held liable as well, so doing a baseline health check is crucial. A new CSM should find out about all liabilities as quickly as possible so they can show that the problems existed before they started.

DISCOVER PAST INCIDENTS AND IDENTIFY FUTURE RISKS

The CSM needs to learn what people within the organisation know about past incidents and get on the same page as them. It's a good idea for the CSM to get the pre-existing liabilities in writing so they don't get blamed for something that predated their joining the company. They should document and analyse any significant past security incidents.

Gathering information about past incidents is fairly easy to do because people tend to remember what's gone wrong. If there were successful cyberattacks in the past, usually those are recorded somewhere, often with risk managers or previous CSMs. Get copies of past emails, meeting notes, or reports discussing the incident. Then work to understand the reasons and root causes that led to the incident. After learning of prior incidents, the CSM should inform senior management or the owners of the company in writing. At that point, it's no longer the CSM's liability.

The CSM also needs to find out what people inside the company don't yet know. They must collect information from internal and external sources to find out if there are ongoing risks of potential breaches and to understand the magnitude of any vulnerability. Gathering this information can be more of a challenge, as external breaches may have gone unnoticed inside the company. For example, if a SaaS provider has been breached, the company may have no idea. There are tools available for monitoring this; the CSM should use them.

DATA LIABILITY AND THE CLOUD

Companies want to collect data, but the more they collect, the more liabilities they take on. Data collection is cheap to do, and just about every company does it; unfortunately, many don't realise that all of that accumulating data can become a liability. The more data that exists, the more data that can be leaked.

NEED TO KNOW

Get the best information available about the exposure your company has. Know and give the right information to the right people at the right time.

For example, think about fifteen years ago—how many systems had your credit card information? Probably one or maybe even none. Possibly just the bank that issued the credit card. Now credit monitoring companies, advertisers, and aggregators collect and sell your information. Data collection has always been around, but there is so much more data today, and it's stored in many more places. Everybody is collecting information, even down to individuals' shopping behaviour. Customer data is valuable. Many companies recognise that value and want to mine it, but they should be careful because the repositories become interesting targets.

For example, PCI DSS standard dictates that cardholder data can't be stored if there is no business reason to do so. Right after the credit card transaction has been processed, the company must remove that data from their systems or make the card numbers unreadable by other means. Not storing the data offers the strongest protection.

Simply put, the more information a company collects and stores, the more liability they have. It's not just Facebook. Any company-collected data that resides in the cloud or with external vendors creates even more liability. Along with that comes a responsibility to protect that data. Personal data should require consent and a valid reason to store it.

LEGAL RAMIFICATIONS

There are significant legal repercussions for failing to store data securely. For example, let's say a financial institution uses an external service provider and that vendor gets breached and loses 100 million credit card numbers. The law in most Western countries mandates that legal responsibility falls on many shoulders: the merchant who lost the numbers, the bank that issued the card, the processor who sent the transaction, the vendors who built their IT systems, and so on. Determining who is at fault may take some time because the liability falls on many shoulders.

For example, in the United States, it's mandatory to provide an identity theft monitoring service for customers who are affected by the breach. Those services aren't cheap, and if you multiply that cost by 100 million customers, you'll see how a single breach can cost a significant amount of money.

In Europe, a new regulation called the General Data Protection Regulation (GDPR) mandates strict procedures for securing customer data. It also provides for massive fines if companies don't comply. The fines for repeat offenders can easily be in the tens of millions of dollars. The GDPR is a regulation with sharp teeth.

The legal requirements for securing data are increasing all over the world; companies that fail to protect their customers' information will see their liability grow, too, especially if they are careless with data. The risks are high right now—regulators are eager to get their first big case just to make an example out of a company. They know if they handle the first company too leniently, they'll have to do the same for everybody, so we expect judgments to be heavy-handed and penalties to be stiff.

CHAPTER FIVE

CREATE A CYBERSECURITY DEVELOPMENT PLAN

As soon as the CSM has a firm understanding of the structure of the company and the current security liabilities, the next step is to create an effective cybersecurity development plan. This plan will provide a blueprint of actions to help them move forward in securing the company's data. This planning stage is when the CSM starts to map out what should be done in the next year to eighteen months.

The recommendations that go into the cybersecurity development plan are based on the real risks uncovered by talking to the stakeholders and leaders at the company. That may be a long list, so an important function of the cybersecurity development plan is to prioritise to-do items. For example, after talking to the stakeholders, the CSM might learn that there are actually two glaring vulnerabilities that have to be rectified immediately based on business needs. Those would go at the top of the list.

In a good case, the requirements will be both compliance- and risk-based. Hence, they make total sense to the company management and also tackle compliance problems. It's a win-win. A risk-based decision might be as simple as purchasing a burglar alarm system. Perhaps the company doesn't have a burglar alarm system in place, yet they're located in an area with a high risk of burglary. Even if buying an alarm system costs tens of thousands of dollars, it could be a smart decision to minimise risk.

Compliance-based items will probably revolve around regulations or standards. Perhaps the business is required to have a business continuity plan and incident response plan in place. For a smaller business, these types of plans usually don't make much sense, so most small companies don't have them. But if it is mandated by a regulation, the small business will have to spend money on creating those plans just to be in compliance, even though management may feel it's unnecessary.

The challenging thing about compliance is that you have to cover every part of the standard set forth in the regulations, one item at a time, until you're done. For most items that are not completed, the CSM must mark them as to-dos and then find the most cost-effective way to accomplish them. That often creates another long list of to-do items for the CSM.

In other circumstances, compliance items can be marked as not applicable. It's a grave mistake to try to meet every single requirement as is. The CSM should investigate whether some requirements actually pose any significant risk or not. Under some compliance schemes, it's quite possible to not implement requirements if there's no accompanying risk whatsoever. This is one kind of an outscoping tactic that professionals use to reduce their scope of requirements.

COLOUR-CODED PROGRESS

Creating a cybersecurity development plan isn't necessarily complicated, though. It could be as simple as making a list of prioritised and scheduled tasks in an Excel spreadsheet. To keep it simple, the CSM can take a Payment Card Industry Data Security Standard (PCI DSS) or an ISO 27001 standard and copy and paste all of the headers into Excel. Then, under each header, they can enter some subtitles for each of the requirements. That would be an easy way to get started.

If the CSM *must* do something on that list, that's considered a mandatory item and would be a top priority. We suggest marking whether each task is accomplished or not using the green, yellow, and red highlighting system in Excel. Once you're finished colour-coding this to-do list, it will become a large spreadsheet full of green and red markings denoting items that are either done or not done, or not applicable, if the items can be outscoped.

MAKING A LIST

Auditors may assume applicability of all requirements, but you can show them the exceptions. Simply create an Excel spreadsheet and list the requirements. Next to each requirement, specify the reasons there is no risk involved.

For example, if the requirement calls for securing the loading bay and you don't have one, you can reasonably opt out.

Give this list to the auditor so they have something to walk away with.

Every task that's marked as done requires evidence to prove that it's been completed. This means you need written, documentary proof. If you have a security control in place but you have no documentary evidence to prove it, then you're not compliant until such evidence can be presented either in writing or by means of observing. It is customary for auditors to first look for documentation. Most often, they will be satisfied with written evidence.

If there's no such evidence, they might opt to observe how the requirement is met or not. That is slower, though. Many to-do items can be brought into compliance simply by writing a short explanation about how you met a certain requirement and where the evidence is. The written documentation doesn't have to be exhaustive; not everything needs to be on paper.

After going through this list, anything that remains unfinished despite your best efforts will become your ongoing to-do list. These will be the items you focus on over the next twelve to eighteen months. After this list is prepared and updated, the next thing to establish is your time schedule and who is responsible for which tasks.

Each of the requirements on the list needs to have a person's name attached to it and needs to be discussed with the person it's assigned to. They're not going to do it if they don't know that they have to. This requires many meetings and a lot of discussion and presentations; it can take a long time to achieve. Usually the cause for delays is not the amount of work to be done but the amount of communication and the number of meetings required.

For instance, if the plan calls for installing antivirus software on the Windows servers, the CSM is probably not the person to do that. That's the IT team's domain. The CSM should not be the one buying the licences, installing them on the servers, checking that everything works and troubleshooting as needed. The CSM should talk with the head of IT and let them know there are requirements that need to be met and then let the IT department do it.

How fast a CSM can accomplish the items on the to-do list and reach compliance largely depends on how fast they can schedule the required meetings and how well they communicate with the people involved, motivate them, and get them started. It usually takes months because just getting a meeting scheduled with any busy business person can take weeks. So the CSM should always be realistic about time frames.

Once the CSM is making progress accomplishing to-do-list items, it's time to look at the bigger picture of security at the company. Now the CSM should think about long-term strategic planning; it's an opportunity to design an effective security organisation within the company and to figure out what long-term budgeting is needed. Do you need a team of specialists working for you? Do you want to apply for a budget to hire some people for new security roles? Hiring permanent positions might be necessary in a bigger company. We'll talk more about budgeting in the next chapter.

PLANNING PITFALLS

There are a couple of common pitfalls CSMs face when creating a plan for the short- to medium-term, like the twelve- to eighteen-month plan we've been talking about. First, too many CSMs use compliance standards as a plan or as the basis for a plan. The potential pitfall here is that the bulk of the CSM's time will be directed toward achieving compliance status instead of working on actual security improvements. If the CSM decides that they need to proceed, from a compliance standpoint, they're going to spend a lot of time looking at the specific requirements to make sure they are met, when, in reality, they can probably comply with most of the requirements just by collecting existing evidence and documentation without actually doing much new security work.

Ironically, a compliance-based plan can actually backfire. If management sees that the CSM doesn't appear to have any to-do items left on her list, they may pull funding of the CSM position. In other words, doing just compliance will get the CSM into trouble. Compliance must be part of the cybersecurity development plan, but the CSM should not focus solely on compliance.

Starting with a purely risk-based approach can get the CSM into trouble too. It might not come back to bite the CSM immediately, but sooner or later the compliance items have to be covered; they cannot be ignored. Every company in a given industry has to do the same things, and other companies will not be happy if their competitor seeks an advantage by not complying with industry regulations. They'll think the company is playing dirty, and they will complain. Then regulators will respond by taking away the company's licence to operate in that market.

If a company has been doing good work with security management based on risks, it will not mean they are compliant. They might be relatively secure but still not meet the mandatory requirements. Being secure but not compliant means the customers may be well protected, but there can be no reliance on this! It means that the company can't put 'secured by' logos on their customer pages, they can't use compliance in their marketing and they will be asked about it often. They will see more auditors visiting due to non-compliance. it becomes a validation problem of sorts. Compliance often leads to certification, which is one type of external independent validation. This helps third-party stakeholders like customers to quickly deduce whether the company can be trusted or not. If they have a 'secured by' logo, then why not buy online?

Neither risk nor compliance can be ignored.

CHAPTER SIX

KNOW THE BUDGET

When the CSM has created a solid cybersecurity plan, after meeting with department heads and senior executives, putting together a current assessment of risks, and identifying compliance issues, he comes away with a detailed to-do list that can move the company forward on security. The work is not done, though. He has one more important step to take—he needs to sell that plan to company leadership.

The highest level of management usually sees cybersecurity as a necessary evil, a cost centre, not a profit centre. Security doesn't produce any services or products by itself, and it makes no profit. So the CSM will have to sell the CEO on the idea that security makes life easier, faster, and more efficient, and hence, it does contribute to the bottom line.

To sell it, the CSM will need to go back to the decision-makers with that plan and explain to them the most urgent items that need to be fixed—not all of them, just the critical ones. And he'll need to know how much it will cost.

When a person first starts as a CSM, they usually don't have a budget in place, unless they are replacing a CSM who used to work there. Many CEOs and C-level executives view security as a cost that should be kept to a minimum. It's not that they are careless, they just don't know much about security. They have their own work to manage. So it's to the CSM's advantage to have the cybersecurity development plan in place first, with the business risks and compliance requirements spelled out, before asking for funding.

Almost every security plan is going to require some level of budget. Ideally, the CSM's budget will come from one batch of money approved by the CEO or CFO—or maybe even another C-level executive who has a lot of money in their department's budget. If the CSM can't get one chunk of money approved all at once by the CFO, they'll need to go through each of the department heads in the company and request to have a portion of the security budget built in to their department budget. This is more cumbersome but works fine.

LOOKING BEYOND THE FISCAL YEAR

If a CSM develops a security improvement programme for the company that takes care of the biggest risks and compliance issues, and they get a budget for it, then what? What happens next year when they've completed the items on their plan? If they don't think about this and plan for it, the CSM might see their budget cut, maybe to zero.

Many companies have a budgeting cycle of one year, but the smart CSM will plan for the long term, not just the next twelve to eighteen months. With security, there are constant recurring costs that are necessary for maintaining every item on the plan. Security measures can't always wait a year for budget approval.

For example, let's say the CSM buys more firewalls or intrusion-detection technology. They need to run constantly, and somebody has to be looking after them, so that requires funding. You must either buy an outsourced service or hire somebody to take care of it. Either way, it's going to be an ongoing cost, which requires ongoing budget, on top of the CSM's own salary. After one year without a budget, someone loses a job or the security controls are left to rot all alone.

Think of it this way: a security budget needs to have a long tail. Security awareness training needs to happen every year and will require a budget every year. And costs go up every year. If you don't get anything done for a year because you don't have the budget—you can't order audits, can't implement new controls, you can't buy technology or services—what's left? Your own working hours. Without the budget, a CSM can rarely get anything substantial done. Failing to plan ahead for budgeting means failing to implement the next year.

BUDGET BENCHMARKS

Most CEOs and management teams have no idea how much they should be spending on cybersecurity. They don't really understand why it costs so much money. So a good place to start is with benchmarks.

CSMs can find benchmarks in studies that describe different types of businesses and how much they spend on cybersecurity. *CEO Magazine* and Gartner, for instance, have published reports on this topic, and they suggest some ballpark budget numbers:

How much should organisations spend on cybersecurity? Cyber attacks and data breaches are becoming so common that all organisations are likely to be hit at some point. Some organisations might use this as an excuse not to invest in cybersecurity, but it's possible to reduce the risk and subsequent damage of an attack—and that option is a lot more affordable than waiting until it's too late.

For example, Maersk announced that it lost up to $300 million (about £225 million) after it was hit by NotPetya, and it still had to deal with the consequences of the attack and upgrade its security measures. Granted, few breaches are as damaging, but the average company is still devastated by cyber attacks. Ponemon Institute's 2017 Cost of Data Breach Study found that UK organisations lose £2.48 million on average after a data breach.

With organisations already investing heavily in cybersecurity—Gartner predicting that global cybersecurity spending will rise to $90 billion (about £68 billion) in 2017—the answer isn't simply to invest even more money. So, what should organisations do?[2]

On average, organisations spend 5.6 percent of their overall IT budget on security and risk management, according to another report from Gartner.[3] But as SANS points out below, most cybersecurity items won't appear under one security budget. They are spread across different budgets

2   Luke Irwin, "How Much Should Organizations Spend on Cyber Security?" *IT Governance Blog*, November 13, 2017, https://www.itgovernance.co.uk/blog/how-much-should-organisations-spend-on-cyber-security/.

3   "Gartner Says Many Organizations Falsely Equate IT Security Spending with Maturity," *Gartner Press Releases* December 9, 2016, https://www.gartner.com/en/newsroom/press-releases/2016-12-09-gartner-says-many-organizations-falsely-equate-it-security-spending-with-maturity.

and are hard to track. The truth is likely higher than 5 to 6 percent. As an example, some CSMs ask for a budget of 11 percent of the IT budget. That seems like a lot of money, but most large businesses with a few thousand employees will pay a full-time cybersecurity manager. They also buy external services and licensed software. That means the 11 percent figure is not far off from real-world budgets. (The trend recently in the industry is that clients are planning to spend even more than that 11 percent on cybersecurity in the near future.) Once you have set an expectation, it's easier to come up with a plan and associated cost. Then if you come in under budget, that will make you quite popular with the CFO.

On the other hand, some companies simply refuse to spend money on security. One company in Singapore had a turnover of about $50 million, but their entire IT budget was only $14,000 a year. They used a government grant to fund it. The IT director wanted to buy licences for antivirus and a firewall and that was it. In a case like that, it's impossible to tell them to budget $50,000 for security.

In most cases, the IT budget provides the most useful comparison because most people think that cybersecurity is about IT. Knowing the IT budget will be helpful for the CSM in determining the right amount of spending on security. SANS states that 'most organisations fold their security budgets and spending into another cost centre, whether IT (48 percent), general operations (19 percent) or compliance (4 percent), where security budget and cost line items are combined with other related factors. Only 23 percent track security budgets and costs as its own cost centre. SANS makes an astute observation which may account for the shortfall in IT spending projections by some researchers and analysts.'[4]

AVOID BEING THE SCAPEGOAT

Selling the cybersecurity plan to higher management is crucial—every CSM needs this high-level support to get things done. This need cannot be overstated. Without leadership buy-in, the CSM will find it difficult, if not impossible, to control the biggest cyber risks in the company, let alone put a robust security programme in place. They may have trouble even implementing the programmes that are mandatory by law.

If the CSM takes their plan to the CEO and doesn't get any support or sign-off for the action items listed there (and the associated cost), the CSM might as well resign. He cannot get his job done. To make matters worse, if something goes wrong and there is an attack, the CSM will still be blamed for it.

---

4   Steve Morgan, "Cybersecurity Ventures Predicts Global Cybersecurity Spending will Exceed $1 Trillion from 2017 to 2021," *2018 Cybersecurity Market Report*, May 3, 2017. https:// cybersecurityventures.com/cybersecurity-market-report/.

The CSM isn't the boss, but the things they do affect a lot of people in the company, so it's a role that's easily turned into a scapegoat. CSMs have been fired because they don't have management support. Something goes wrong, and people think, 'I didn't like this guy anyway, and now's my chance to get rid of him. I'll blame him.'

In one instance we learned about, a CSM was friends with the managing director of a government organisation, and they had a strong working relationship. But a small team was of the opinion that the managing director was no good and needed to resign, so they burned the CSM for some minor non-compliance issues and used that as a tool against the managing director. Political games played in security and compliance can get dirty. Nobody wants to be the scapegoat. This is another reason the CSM has to build a network of support within the company.

Compromise can be key to gaining support. If senior management doesn't want to spend money on security, the CSM might negotiate a modified plan. Can we do half of the security plan this year and half next year, to lower the spending? Can we use more internal resources for the different tasks? Of course, there will always be a cost for internal resources, but it's often possible to haggle about the price. If the CEO or CFO is willing to discuss the plan, that's an indication they want to do it, but maybe not exactly as proposed and maybe not all at once. They might want to understand it first, then make some adjustments. That's a good sign.

Communicating the plan and the budget is the first and biggest test of management support for the CSM. It's not the first step, though. The CSM should have already cultivated a high level of visibility with senior management and used it to start talking about the necessary changes long before presenting the plan so that the senior leaders won't be surprised, especially the CEO.

Management support is crucial for anything related to security—not just the budget. The CSM simply cannot be successful without management support.

BUDGETING WORKAROUNDS

Sometimes the CSM won't have a centralised budget or a large enough budget to pay for everything, so they need to go to someone—perhaps the CFO—to ask for money. Of course, that means they have to know that person well enough to talk about his budget and spending and what needs to be done with security.

It pays to play the C-level money game—make it your business to be friends with the CFO. Every year, a few months before budgeting, all the C-level executives and department heads try to get close to the CFO of the company, or whoever has final approval on spending. The CFO is usually the most powerful person within those department heads. She might not make the decision alone, but she has powerful influence within the institution. If the CSM gets a budget assigned by the CFO and is already friends with her, it's going to help them immensely.

As long as the CFO understands the budget and thinks it's a necessary expense for the company, it's unlikely she's going to cut that budget later. If the CFO is sceptical of the CSM's budget numbers or doesn't understand the need, there's a risk the CSM will be laid off or the CFO will find other reasons to cut the budget.

The CSM can prepare for some of these possibilities. When building a team, the CSM should think like an executive—when a company has a good year, every executive hires staff into less-senior roles, just to be able to cut those positions when the next layoffs come. People can shift into other roles, and there is plenty of headcount to let go when layoffs come. Similarly, a security management team of five people—which would only happen in a big company—can always consider bringing on a sixth person, just in case they have to shed numbers in a future budget cut.

There are many tricks and tactics like this that a CSM can use to get a fair budget approved, but the politics and approval processes in each organisation are different, so the most important skill for a CSM is learning who controls the levers of power. Then they can figure out how to operate those levers to maximise security and minimise risk.

PART II

COMMUNICATION

WORKING WITH EACH TEAM TO CREATE CHANGE

In this part of the book, we will explore some of the different types of teams within a company and consider how to effectively communicate with them. The needs and requirements will be different between departments and among the people who work in different roles. Understanding these differences is necessary for a CSM.

Our main focus will be to understand the motives and roles that each function may have. The functions we'll cover include risk management, human resources, IT, legal, finance, facilities, and operations.

CHAPTER SEVEN

RISK MANAGEMENT

Securing a budget is one of the most important goals for the CSM, but the numbers will be meaningless if the company doesn't understand the risks the cybersecurity plan seeks to mitigate. Almost anything the CSM wants to get done requires them to demonstrate an authentic risk. If the CSM can demonstrate the existence of real vulnerabilities within the organisation, that will motivate leaders to take action.

TALKING TO PEOPLE ABOUT RISK

It's not just about the numbers, though. Risk management is actually the fine art of giving people difficult feedback but doing it gently and professionally without causing alarm. Here's an analogy: if you are a doctor, it's unprofessional and callous to tell the patient bluntly, 'You have a deadly disease, and you might not survive it, but it could be curable.' An effective physician will explain the situation honestly but gently, with a touch of optimism. He might say, 'You have one mole that concerns me on your left shoulder. It seems to be limited in size, so the probability that we can get rid of it is fairly high. I'm not making promises, but you have a good chance of a complete recovery.'

Similarly, the CSM can't just tell the CEO, 'Your company is in trouble.' The CEO will immediately start thinking about the worst-case scenario. A panic-based response is seldom helpful. As with the doctor and patient, it is much better to explain the risks while providing possibilities for resolution. This is the best framework to use when discussing risk with senior management.

RISK OWNERSHIP

In most companies, some people understand risk management and others don't. In some, nobody knows anything about it. Even if they talk about 'risk management', they might mean something else, like threat or danger. If they don't own the concept of risk, they probably won't take action to mitigate risks.

For example, we recently learned about a company that was hacked, and their passwords were stolen. We reached out to the company to ask if they wanted to buy an assessment to find out what else had been stolen and how to remedy the situation. They agreed that was extremely important, so we thought we were moving forward.

Not so. They said that they were not going to hire us or anyone else. Of course that's ridiculous, because the passwords were out there. The risk was real; someone was robbing the company right then. But since no one owned the risk, nobody did anything.

Who should own risk? Not the people you might expect, like the risk manager, security manager, or fraud expert. Instead, the risk owner is someone who can assign a budget to address it. If they can't assign a budget to make sure the problem is fixed, they cannot be an owner. The owner of a risk is never someone without a budget and authority.

CSMs should know who is responsible for every risk. The owner should be able to choose what they're going to do with the risk, the measures they will take, and how they're going to pay for it. Because risk is always tied to an operation, it is closely connected to profit and opportunity. Risk and opportunity go together. If you ask who owns an opportunity, you'll usually find the owner of the associated risk.

FINDING THE OWNERS OF RISK

Someone needs to be in charge of risk management at all levels and in each department. This isn't to say that all businesses should have a separate risk management role, but someone in each department should take responsibility and ownership of certain risks and levels of the organisation.

In HR, somebody has to be responsible for HR risks. That's mandated by law in many countries. Supervisors can be owners. In Singapore, for instance, construction supervisors must make sure every employee wears a helmet and boots, even if they're just drawing maps on the construction site. If somebody were injured at the construction site because the supervisor didn't enforce the regulations, he could be arrested or fined.

Similarly, in the IT department, the CIO, who assigns duties and tasks to the people who manage malware or ransomware risks, must acknowledge that IT risks are in their domain. This is something the cybersecurity manager has to be aware of too. It also helps if the CEO is aware that risks are part of the business and need to be managed.

COMMUNICATING RISK

CSMs need to map all the risks in a language that management understands, or nothing will happen. The CSM should gauge the audience to find out the right way to talk with them.

There is no right formula for communicating risk because each company is interested in different things. They might be focused on values, public image, profit, or employee morale. Inexperienced CSMs sometimes try scare tactics. They say things like 'There's an unknown risk and you can never be prepared enough, so you should buy more stuff.' That's too vague and unquantifiable to work. The CSM should emphasise the company's specific interests. For instance, many companies only care about profit, so for them, it helps to quantify the risks in terms of potential lost profit. Other companies that are protective of their reputation will listen carefully if the CSM mentions a reputation risk. Each company is different, and a company's values can change even during one year, so the CSM needs to stay on top of this.

One of the soft skills used in risk communication is to keep it impersonal. That way the people who are responsible for the risks don't feel like they're being blamed for them. Risk managers become skilled at discussing risks without making anyone feel personally responsible. Risk management would never say to an IT professional, 'You should do a better job with downtime. Twice a year, we have unexpected downtime and a communications break, and nothing's working. It's your job to keep everything online, and you failed to do so. The cost of your mistake to the company is $200,000.' Placing blame like that would make it very personal and would lead to bad feelings and low morale, and soon, nobody would report risks because they know they will be stabbed in the back. The way to communicate these issues, without making it sound like a personal attack, is to talk about the objectified risk and not the person.

RISK APPETITE

The CSM also needs to assess management's risk appetite. Risk appetite measures how much risk a company is willing to tolerate and what is an acceptable level of risk for them. Professionally led companies traditionally have a bit more risk appetite, while small, family-led firms especially, tend to be risk averse, but it's hard to predict and depends a great deal on the tone of the owners and the board. Defining risk appetite requires a high-level decision in the company; the decision has to be ratified by the board or at least by senior management. The board must take the lead in risk management, and then those decisions must be executed by management. Only then does risk management actually implement it.

RISK PERCEPTION

We talk a lot about quantifying risks, but some risk management decisions are definitely made on an emotional basis. People sense danger or feel threatened. Those reactions are important but difficult to measure. Which risk is bigger: your daughter crossing a busy street or an airplane jet crashing to the ground with your manager on board? On an emotional level, it's definitely your daughter. On a statistical level, maybe not. That's how emotions can get in the way of analytical decision-making.

People make emotional decisions and are notoriously bad at quantifying risk. For example, some people develop irrational fears of a shark attack or a plane crash, when the probability of either happening is extremely low. Even if they know something about impacts, most people don't understand probabilities. They tend to worry about risks with a minute chance of happening and disregard risks with a high likelihood.

The same goes for managers and owners who assess risks in their business. They might be very professional in their primary job function, but they are usually ineffective at assessing risk probability. We've heard many managers say, 'Communication breakdown rarely happens to us and doesn't have much effect.' That's almost never true; if we do the math on the probability, it's actually a big risk.

CHAPTER EIGHT

HUMAN RESOURCES

The human resources department helps the company find people, hire them, train them, fire them, and manage the whole personnel process. For some reason, there are a lot of jokes about HR people being evil, maybe because they control people's careers. But they are essential to any organisation. They have a say in the allocation of money for salaries and benefits. They are the gatekeepers for employment issues and education and training. Because cybersecurity is first and foremost a people problem and only secondarily a technology problem, CSMs should consider HR folks essential partners.

HR actually tends to be quite helpful to the CSM. We don't remember a single instance when HR hasn't wanted to help improve security. In fact, HR managers usually become friends with the CSM.

BEST PRACTICES FOR HR

Advocate for HR to pay attention to basic security measures:

- Review all aspects of the résumé or CV.

- Conduct a background check. (If you use an external service provider for the background check, check on them too.)

- Call previous employers.

- Verify candidate's identity.

- Assign equipment, keys and access rights, and have them sign for these.

- Provide induction and familiarisation training.

The CSM should find those few cybersecurity issues that actually belong to HR or where HR is instrumental, then enlist their help. The CSM can place certain checks and security controls on the employment process, then just stay hands off. Set HR up with the right tools, then let the HR machine run on its own.

SECURITY IN THE HIRING PROCESS

When starting on the job, the CSM should meet with someone in HR as soon as possible—preferably an HR manager or director—and have a discussion about security. Ask the HR manager or director the questions below.

Their answers will help you understand which security measures are in place and which are not.

- What do you do for security when new employees start at the company?
- Is there any formal security training provided?
- How does the company maintain its security awareness continuously?
- What do you do when they leave the company?

The discussion will also help establish a relationship between the CSM and HR personnel, and establish the CSM as someone who is there to help. In fact, if the company is big enough to hire employees with cybersecurity skills, or if it's hiring a cybersecurity specialist, the CSM might even be able to help HR craft the security requirements in their job descriptions. For example, if they are looking to hire an internal auditor for security, HR will probably not be able to define what skillset and capabilities the role and the candidate should have, but the CSM certainly can.

The CSM can guide HR to create—and use—a hiring checklist. The checklist should include simple things like verifying the candidate's identity—are they really who they claim to be? HR should check each new hire's CV and certificates to make sure they match. They should also have procedures in place that require the new employee to sign for any computers, keys and keycards issued to them. Other checklist items include assessing which user access rights are appropriate for the new employee—what access will they get to the network and services, both within the company and externally? The HR manager should also ensure that the employee receives access induction and familiarisation training when they start the job. The CSM should make sure that HR has this hiring checklist in place and that every hiring manager uses it each time someone is hired.

Next, the CSM should make sure HR is providing at least basic security training to each new employee. The company may have internal policies and procedures in place about security, but if they're not part of the induction training, then the new hire won't know about them. If she doesn't know about them, she won't comply, leaving the company open to a possible security breach. The responsibility for that breach would then make it through the chain of responsibility, from the supervisor all the way up to the CEO. Ultimately, the responsibility lies with the company.

Security in hiring requires attention to detail, but at the same time, the CSM should make sure they don't lose track of the bigger picture: usually the biggest risk is hiring someone who is not a good fit for the company. The CSM can't forget that the main goal of the HR process is to find the right person for the job.

SECURITY IN THE FIRING PROCESS

During firing, HR should also have a checklist. The hiring manager or supervisor essentially reverses the hiring process:

- What does the terminated employee have that they need to return?
- Did they give it all back? Did they get a signed receipt for returning their equipment?
- When should account access be disabled, and where did they have access to?
- Are there any foreseeable disputes that make it wise to end access early?
- Or will the departing employee keep temporary access to computers or accounts?

It's up to each company how they want to implement different types of exit scenarios. But there should be a plan and set policies in place. In our experience, which includes firing CEOs, it's best to have an established checklist and follow it every time without fail.

THE POWER OF PUTTING IT ON PAPER

Have your policy and process documents at the ready. People tend to accept them at face value with a less emotional response.

If possible, it's useful to have the employee sign the checklist and agree in writing that the computer will be returned that day before they leave, that they understand that their email password will no longer work after a certain date and so on. When they sign that list and hand over the items, there can be no dispute about what was agreed to and what was done.

The best practice is to make security discussions with new hires a standardised process. If you make it a process, it's not personal. It's just a typical work conversation. 'I have this HR paper; can we go through it together? We have to do it. It's policy.' This is much easier than arguing over a computer or access rights or what to do with an email account. If the employee doesn't like how they're being treated, you can blame the paper, the policy, the process and the bureaucracy.

SECURITY DURING EMPLOYMENT

During their time working for the company, each employee will need ongoing training and education on security issues. HR usually arranges these sessions, so when the CSM needs to deliver a security awareness programme, they should go right to HR for help.

They will probably help the CSM set up a short training course, delivered either as a lecture or as an e-learning experience. The format is not that important, but tracking attendance is. Many CSMs get into a hard spot when auditors ask them to provide attendance logs from the training as proof that awareness training has been provided. These logs can't be fabricated in the aftermath. Is every person in the company actually participating? How often does the training happen? An annual programme is good practice, but does it need to happen more frequently?

Another important question: who gets this training? Typically, companies think that everyone should know basic things related to security, especially anyone who has access to confidential company information, but different specialists have different needs. Maybe IT employees should know a bit more than the average employee about cybersecurity. What about supervisors? Do their additional responsibilities—hiring and firing—mean they need additional security training?

CULTIVATING SECURITY AWARENESS

All of these things have to be part of a cybersecurity awareness programme or plan, but the overall goal is to improve security awareness within the company. There is no fix for human stupidity or carelessness, but security awareness training can help avert the worst problems.

We've spoken with cybersecurity managers in high positions and asked them how much emphasis they put on technology and how much on awareness and human behaviour. One said he's putting half the budget on awareness. It sounds like a lot, but think about what happens if you don't. A simple email can be used to steal millions of dollars from a company. If employees aren't aware of such a risk, no amount of technological protection can save them.

We worked with one person who received such an email. It appeared to be from the CEO and said, 'I'm travelling and this is a busy case. Here's the bill that should be paid to a service provider in the Far East. It's nighttime out there. Please just put the payment in and then we can handle the proper paperwork later.' The bill looked legit. The email looked legit. The only thing that was not correct was the account number and bank. That company lost $17 million because of that email. Security awareness and training is the only thing that helps in those cases.

Even drastic technological measures to prevent cyberattacks can fail. Recently, in Singapore, the government enacted a new regulation that requires all public entities—governments, schools, hospitals—to disconnect from the internet; they have to use different computers for internet and physically separate networks for internal needs. It sounds like this change should help, but people can be fooled into overcoming that 'air gap'. If that happens, this technical measure isn't helpful. The money stolen in the example above would still be gone. Just recently, the health information of 1.5 million people was stolen from SingHealth, a major healthcare provider in Singapore. A simple click on a wrong email could still cost your organisation millions.

Attackers pay attention to human behaviour; they will try to fool somebody within the company into unknowingly assisting them with the fraud or hack they are performing. If all staff remain alert to these attempts, many of them can be avoided. Not long ago, someone tried to get billions of dollars from the Bank of Bangladesh in one of the largest attempted cyberattacks in history. In the end, they were only able to steal tens of millions, not billions, all because someone in the company noticed a typo in one message and put a halt to the scheme.

These examples illustrate the need for security awareness training. Human carelessness is hard to regulate. But training employees at least every year can prevent needless disasters like those described above.

MAKE HR HAPPY

HR is happy when the CSM helps them acquire the proper training and create the guide for the training. But they usually do not like when a CSM forces on them a new security checklist or process. It feels like extra work. The best way to help them gain ownership is to talk with them and let them come up with the process and structure on their own. This way, HR can figure out for themselves the best way to do training and then build a programme from scratch. A CSM who imposes training requirements may face resistance and have to do a lot of pushing and pushing to make progress. But when a CSM is clearly there to help HR implement their own ideas, their input is usually welcome.

BUDGETS FOR TRAINING

Training is going to require a budget, and that money might come from a variety of places. The HR budget will likely cover induction training and basic security training. Extended training, such as a week-long course on security auditing, will probably come from the auditing department's budget.

Say a finance institution needs some software development. They would be required, usually, to train their developers with cybersecurity in their development work. It's fairly technical training. HR would be instrumental in first getting basic training to everybody in the company, but beyond that level, they probably need to involve different departments separately.

Centralised training stays with HR. Specialised training migrates to the various departments.

In the end, a company is nothing but a bunch of people working together. As in any relationship, reciprocity is key. HR, instrumental in a lot of important activities in the company, can cause problems for the CSM if the CSM is not helpful to HR in return. On the other hand, we've never seen an HR director who didn't want to improve cybersecurity. Make HR your friend, and you will be much more successful as a CSM.

CHAPTER NINE

IT

The CSM and IT staff are often natural friends. They usually work closely together because they are both responsible for implementing many of the technical controls. ('Control' means solutions, processes, policies and so on.) Since IT staff are the ones who do the actual technical work, they're generally pretty familiar with security issues and work well with the CSM.

The relationship between CSM and IT staff isn't exactly one to one, though. If you think about it, IT is not just one job; it's many different jobs in a wide range of areas. Big companies usually have tens or even hundreds of people working in the IT department, and each of them specialises in a different area. The CSM works with them all as a security professional with a holistic view of cybersecurity management in the organisation. IT's job is to implement the controls and technologies, and the CSM helps them to discover what's necessary.

As an example, let's say the CSM wants to protect users from malicious links so they won't click on anything risky. IT will probably have to implement some technology, like a web proxy that all computers in the network are using, so that any link they click is verified or scanned while the traffic is flowing through that process. Or maybe they will install a spam and email virus scanner solution that removes suspicious messages altogether. Either way, the CSM needs help from IT to find and implement these solutions.

WORKING AS A TEAM

Not surprisingly, it's hard for the CSM to get things done without the help of IT. Traditionally, IT is intended to create efficiency in a company and save money. In the traditional management view, its purpose is to cut losses or improve performance. That focus on efficiency means IT holds a lot of power when it comes to choosing and implementing new technologies. If the CSM recommends a purchase and the IT department says, 'No, we don't need it', there's a high chance the purchase will be delayed or denied.

Many old-school directors still think IT exists mainly for

efficiency. Newer generations consider it an integral part of business, a tool used to gain competitive advantage. Modern eyes tend to see IT as a strategic asset.

Say a CSM wants to add another firewall to the network, along with new authentication methods. If the proposed change will make processes slower than they were, IT will probably say no. Their mission, remember, is to make things run faster. So it's crucial for the CSM and IT to work together, and try to find agreement on what solutions are necessary. Some CSMs actually work within the IT department.

Whatever their position, the CSM needs to know what's going on in IT—and outside of it. IT today involves more outsourcing than ever before, so the CSM should be sure to follow up with available third-party security services and become knowledgeable about acquiring and managing those services. If someone is running an outsourced mail antivirus scanning service and IT buys that service, the CSM needs to know because they will need to set processes and procedures to maximise the benefit of that service. They'll need to work closely with IT to do that.

The CSM should make friends with the whole IT team, get to know them, and win them over. The CSM should know all of the IT key people by first name—the CIO, the firewall administrator, and everyone else in the IT department they rely on to get things done. It couldn't hurt to also learn their spouses' names. If these relationships aren't close, the CSM won't really be able to work with the IT department effectively. It is well worth the effort for the CSM to build relationships with all levels of the IT department. Gaining the trust of the IT people often requires that the CSM be a relatively tech-savvy person. Sharing the lingo and IT culture often goes hand in hand with sharing technical expertise.

THE BEAUTY OF BONDING

Bonding is the key to getting things done without getting tangled up in bureaucracy.

You can just walk up to people and check in with them. Join their tribe and learn to speak their language, and things go much more smoothly.

CHAPTER TEN

LEGAL

Security concerns can be split into two camps: internal and external. Internal includes all the systems inside the company's walls that the company or organisation has direct control over. Handling internal security is easier than handling external security because it's within the company's domain and control. External is much trickier. External includes outside consultants, outsourced systems, anything in the cloud and all the things that happen outside the walls of the company.

Everything external is managed by agreements, so the CSM needs to work with legal to include legal text in every contract regarding cybersecurity requirements for every outside vendor. It's not enough to just say to an external supplier, 'Can you be more secure? Can you not take any risks with our data?' That is unenforceable. But when these requirements are written into legally binding contracts and service-level agreements, contractors must live up to the terms of the legal agreement that they signed. So security must be clearly spelled out in a provision in every contract with every external supplier.

For a real-world example, we can look at the payment card industry (PCI) standard. Credit card organisations like Visa and Mastercard require that participants adhere to strict security standards, and they expect those members to require the same of their customers, which are the banks or card issuers, payment processors, and merchants. Then those entities have to require the same things from their customers, and so on. So the chain of compliance and requirements goes all the way down to the last merchant who buys a payment terminal in his or her little shop.

The chain of compliance seems logical enough, but in reality, managing the legal requirements can get very complicated very quickly. Take the case of a major real estate company that had historically used contractors and design agencies to build and renovate buildings. After years of operating one way, a new national security standard came along that said, 'You need to require certain things from all the subcontractors who have access to your information, your premises or your facilities.' Suddenly, all of those small suppliers had to start mandating those same security requirements from all of their subcontractors and employees. It was a huge adjustment.

In the construction business, the list of security standards is huge, with hundreds of requirements. When the various subcontractors and agreements with suppliers are figured in, the situation is even more complex. To further complicate things, the people working in this field—construction companies, design agencies and architects—generally have no concept of cybersecurity or security standards.

For this real estate company, handling all the new legal requirements was overwhelming and frustrating. They had to send auditors to all the external companies only to learn that they weren't compliant and probably were not going to be for a long time.

LEGAL AND SECURITY

The CSM and legal staff usually get along nicely, though the legal staff probably don't know much about what the CSM does. The CSM should start by meeting the lawyers in charge of drafting contracts and end user licence agreements to discuss security requirements that should be included within those contracts.

The first thing the CSM discovers in these conversations is that not all attorneys understand security. Usually, lawyers and legal advisors specialise in one area of law, like contracts and IPR. But if you ask them about defects in software and vulnerabilities, they have no idea.

This becomes a teaching opportunity for the CSM. He might be able to provide the legal team with invaluable support, which they will appreciate. If they have a collaborative working relationship, the CSM might see the legal staff coming over to ask for help when they are drafting new agreements for security or compliance.

That's the perfect time to try to influence the contracts that will govern relationships with outside service providers and suppliers—before the agreements are signed. It's usually very difficult to fix problems after the agreements are signed. There's always a cost involved. It's critical that the CSM gets involved in all the new acquisitions and service provider agreements as early as possible.

Whatever the CSM and legal do together usually involves creating some sort of contract, as well as a standardised template with set security clauses for different uses. A typical case they might work on together is buying data centre services. The contract should clearly define operational security and services in the agreement when they buy the service. Some of the language can be templated and some of it may require customisation, but it's a good idea to include right-to-audit to all the templates.

CONTRACTS THROUGH THE SUPPLY CHAIN

There is not a single company in the world that doesn't

use some sort of supply chain. Employees have to have cell phones, an ISP, a SIM card, and an agreement to provide a voice service—otherwise, their salespeople cannot make calls. That's a simple supply chain. But some supply chains are huge. One logistics company we know of used to have ten thousand suppliers for transporting cargo. Other types of supply chains, especially in technology, are extremely complex. Consider all the parts and suppliers that go into making one iPhone.

Most companies work with a variety of suppliers, then those suppliers each use suppliers, and so on. It can be tough to track the supply chain all the way back to the source. Most supply chains are so complex that it's not realistic for any cybersecurity manager to expect to implement security standards for the whole supply chain from start to finish. In reality, the CSM has to focus on just the few that matter most.

Working with the legal team, the CSM should focus on the biggest suppliers and audit those agreements. Typically, supply chain contracts last for a few years—at least one to three years. So when each agreement nears its end date, the legal team and CSM should work together to see if there's anything they want to change or add to that agreement. With a three-year deal with a service provider, there is three years' time to think about what needs to be written into the new contract regarding security.

MAKING LEGAL HAPPY

Clearly, the CSM will benefit from a healthy relationship with legal. Legal will too. The lawyers don't know the security area very well, so the experienced CSM might be able to insert a few beneficial tricks into their agreements and agreement templates. Security clauses that make their way into templates will be used across many agreements. The legal team will also be grateful if the CSM helps them clear out some requirements from contracts that shouldn't be in there, clauses that create unnecessary liabilities.

For example, in software development contracts, there's almost always a clause about fixing significant flaws in the software. If the CSM and legal agree that a security vulnerability constitutes a significant flaw, they can put a clause into the contract saying that all vulnerabilities are significant flaws that should be fixed by the service provider. If legal doesn't have that definition in the contract, it will lead to a negotiation every time there's a vulnerability found in the software. Who will fix it? Who will pay the cost? An experienced CSM can help legal a lot in this regard.

It pays to keep the security language in contracts as beneficial as possible. For example, when purchasing software that's supposed to be secure, the CSM should try to write into the contract that they can conduct third-party audits.

If the audit fails, the contract should stipulate that the vendor has to fix it at their own expense.

So much goes into cybersecurity. The CSM is working within the organisation to create change, and legal can be a great partner in achieving that goal. Every CSM must learn to work closely with the legal team.

Lawyers are not only about contracts, though. They also provide support and insight. If the CSM fosters a good relationship with legal, they can rest assured that if they get challenged, they can depend on legal to help defend them.

CHAPTER ELEVEN

FINANCE

Because every move the CSM makes must be supported by resources, the topic of funding is never very far from a CSM's mind. They will want to get to know the most powerful person in finance, the CFO, or chief financial officer. CFOs have tremendous influence over how effective a CSM can be in her job. The rest of the finance team is usually made up of financial controllers and analysts, and while their roles are important, the CFO is the key decision-maker.

The CFO is sitting on top of the treasure chest of the company, and everybody comes to the CFO at least once a year asking for money, especially if they need additional budget over what they had last year. Everybody at the top is competing for their share of the treasure chest, including the CSM.

SPEAK THE LANGUAGE

It may seem like the CSM and CFO speak different languages. CFOs understand money and budgets really well, but maybe not security issues. Cybersecurity managers who can translate their needs into financial language—business continuity, revenue streams, profitability, etc.—have a better chance of receiving funding. The CFO appreciates it when the cybersecurity manager explains how they can increase profit and reduce fraud by using proper security controls. When possible, the CSM should quantify the payoff with numbers that demonstrate the effect security efforts have on the bottom line. Any CSM who can do that will find friends in the finance department.

Let's say the CSM is trying to communicate the presence of a business interruption risk. If the data connectivity to the data centre goes down, it will have certain impacts—the customers will be angry, and the company will have to spend a lot of money on recovery. If the CSM can demonstrate this will cost $200,000 in all, it makes sense to invest $100,000 this year for the second communications line to make sure everything works in case of a blackout. A CFO will understand that logic. The CSM needs to make the proposition clear, telling the CFO, 'If you approve the budget, this will be your outcome, and if you don't, this could be your outcome.' Because the CFO thinks in numbers, with this simple example, the CSM can demonstrate a clear return on investment.

Another way to frame the importance of security measures is that they protect the company's investment. If the CFO is considering how many people to enlist in cyber-security to make sure investments to date are protected, how will he decide? The CSM can convert the numbers for them so they understand the risk. Say the company is worth $1 billion; maybe spending $1 million or even $100 million would be reasonable. The CSM's job is to show the CFO, 'If you put this much money in, we can be sure that we will wake up tomorrow to the same company and the same assets.' Asset protection language usually works for CFOs, owners, and boards; their basic rule is: don't lose money. Security is not just about loss prevention; it's a way to preserve assets. Owners will be interested in this.

The risks are real, but too many organisations ignore them. We once worked with a government organisation that had a financial process to pay millions in tax revenue back to the states a few times a year, kind of like a tax refund. They had known all along that there was no authentication between the systems that executed these huge dollar transactions. In this case, it was possible to change the account numbers of senders and recipients in the payment system without authorisation, so an insider could have signed in, changed the recipient's account number on a $100 million transaction, and gotten away with it. There were serious risks that people on the inside could collude to move the money somewhere else—into a dark offshore account.

The government organisation was aware of the risks. They acknowledged the risk, and they talked to the CSM, but they didn't do anything about it. Billions of dollars in transactions were at risk. It's baffling to think these kinds of risks remain on the table with potentially huge implications, and a single individual, like a CFO or a director, can just accept that risk as residual and be okay with it.

Companies, and people, don't always act rationally, and that's as true of CEOs, CFOs and CIOs as of everyone else. Every CSM will learn this lesson over time. Risk acceptance will always stay with senior management. Knowing that, the CSM will be able to sleep comfortably.

CHAPTER TWELVE

FACILITIES

We love it when companies start focusing on cybersecurity. Too often, though, the new focus distracts companies from basic security measures that have to be taken care of, no matter what, like physical security. Physical security may seem too basic to spend much time on, but it forms the foundation on which the company, and the CSM, must build all security.

We have seen too many businesses post security guards in the lobby and implement access control systems yet fail to train front desk staff to adequately verify the identity of anyone entering the building. It's often too easy for an intruder to give the front desk officer a fake business card in the lobby, ask for access to a certain floor, and head on up. Or they might use the fire exit or the loading bay to gain access to areas that are supposed to be secure. We see it all the time—doors on the way to the most secure areas are left open and unlocked, and nobody is checking badges. It makes no sense, but it's commonplace.

The facilities department should be the team in charge of all of this—not only the physical building and all of the security technology but also the related services—the lobby, receptionists, guards—and the access control technology, such as the burglar alarm system. Facilities concerns are directly related to cybersecurity because all of the computers, switches, and networks within facilities are usually connected, so they are key to securing all facilities as well as systems and suppliers.

Systems put in place to secure facilities are often vulnerable in surprising ways. In Saudi Arabia, for example, a huge company installed a new access control system that opened and closed their doors. This system was connected to the internal network of the company. Their computers and systems were sitting on the same LAN. When they hired us to test whether their system was secure, the first thing we did was to scan the system to see what was there. At that instant, all the doors in the whole building locked. No one could get in or out; this was not the type of security they were looking for. The bottom line is that physical security solutions are seldom tested for security; meanwhile, everybody is selling security technology. Some of it is secure, but some is awful.

FACILITY SECURITY = IT SYSTEMS

These days, facility security systems are IT systems, the same as everything else.

CCTV, access, HVAC, alarms, the IoT (internet of things)... lots of small computers that usually get little attention. Facilities buy them, make sure they're running and working, but then they forget about maintenance. Meanwhile, IT doesn't usually have access because they didn't buy them.

Facility security is the Achilles heel of many companies today.

Even worse, the company was using a wireless WEP security, which has been known to be insecure for a long time; it can be cracked in minutes or even in seconds. Anybody could have hacked into their wireless network and locked the doors. On top of that, the lock function itself wasn't foolproof; they discovered that it was possible to force the doors open as well.

Firewalls are like the gate to your network; but what's the point of having a firewall protecting your servers if you don't have a working lock on the door? Every company needs both. Even with locking doors, cybersecurity is critical for physical protection of all the assets on the premises.

OVERLAP IN SECURITY

Physical security and cybersecurity overlap. Even in the era of cloud-everything and wireless-everything, there is still a need for wired networks. The cybersecurity manager needs to interact with the facilities people to make the right choices for protection of their physical systems. For instance, access control systems should be segregated away from workstation networks. All of those little closets and server rooms with switches, routers, firewalls and cabling should be well protected. Unfortunately, facilities people usually don't understand the connection of facilities and cybersecurity unless the CSM helps them understand.

For example, a luxury hotel in Switzerland switched from physical keys to completely electric locks. Guests had an access card or a token and could beep their way into their rooms. It worked well until a hacker came along and locked all the guests out of their rooms. It was a big scandal in the media; the hotel had to remove all the networked locks and replace them with old-school locks with keys. When facilities span more than one building, the challenge is even bigger. Access control systems are usually implemented per facility or per building installation. If you have ten buildings, it's a nightmare to manage. If you can put the control in the cloud, however, and treat all those facilities as one entity that's managed centrally, life gets much simpler. Once people started doing that, cloud-based applications and cloud computing companies popped up everywhere, offering this service. Those cloud providers are now using that cloud technology to control security access to a lot of different customers and buildings all over. Of course, this brings new risks. If somebody breaches the security of that cloud, they can get access to any of those clients and facilities.

We've also seen some other hacks with cars that are wirelessly connected, such as when a thief opens a car's locks, steals the car, and sells it for money. It's an emerging problem. In the future, it's likely that we will see more of these car-related cyberattacks.

WORKING WITH FACILITIES

Facilities people might be office managers, or they might have the responsibility of acquiring janitorial services, landscapers, and receptionists. They may have a role in the maintenance of the buildings and facilities. If there is a water leak somewhere, they will be the ones to figure out how to fix it or to find someone who can. Sometimes they manage the security control technologies in the buildings. Put simply, they manage physical building assets. The reason physical security rests with these people is that they often buy the security systems when a building is first constructed. Facilities people now need to buy cybersecurity technologies, too, because everything from access control systems to CCTV are all controlled by computers.

If the CSM is dealing with a facilities team who is not security savvy, they need to help team members understand what security is all about. The facilities people might not even realise they need security. If they don't, the CSM has to sell that idea or help them discover it. The CSM could ask, 'What happens if someone hacks the internal network, then hacks the access control system? What might happen?' Faced with this question, facilities staff might realise it would mean someone could open all the doors or flood the floors with water, causing major damage. There's also a risk of explosions, contamination or other bad results. This is a deadly serious concern in production plants, especially for medical or chemical manufacturing.

We've seen dozens and dozens of server rooms that don't have any environmental controls for cooling and moisture removal, or redundant power or drainage. Or maybe they had controls, but they were really shoddy or didn't include a redundant cooling system, where a backup system takes over if the primary system fails. In a server room that's running a critical business operation, maybe even a big plant, with nothing to secure it, what happens if the cooling breaks down in the middle of the night? The next morning, it's 42°C in the server room, and most of the servers have shut down to protect themselves. Now there's a production break and a disaster all because facilities didn't know how to physically equip the room in the correct way.

SERVER ROOM SECURITY

When a company has server rooms or data centres, the CSM should make sure these security controls are in place:

- Redundant cooling

- Uninterruptible power supply (UPS)

- Drainage

- Raised floor

- Cleanliness (no dust or moisture)

Big data centres usually have standards for this.

Most data centre service providers have some measures in place but no agreement about maintenance or testing. They might have implemented UPS—uninterruptable power supply—systems, but nobody is maintaining them. Or they might have rooms full of batteries to endure an hour of power outage but nobody maintaining or testing them, so it's never clear if they're working or not.

Some measures are simply misguided—yes, sprinklers are very common in server rooms, though they are a terrible choice for mission-critical server rooms because they will destroy the equipment as surely as fire will. What's more, the systems themselves can fail and pose a risk to the equipment. We have also seen highly flammable liquids, piles of combustible materials, and hosts of drain pipes dripping from the ceilings. These are not just in-house errors; even midsized and large hosting service providers make these mistakes all the time.

The CSM has to know about these physical risks. The CSM might not be the best person to solve the problems—that's the facilities team's role—but he has to communicate that there is a risk and that something should be done about it.

If there are a number of physical security problems that should be handled by the facilities team, the CSM should help facilities people to do a risk analysis that includes all the things that are a risk or a threat to the company, assess the likelihood and impact of each, then make a plan for which ones to fix first and find the budget and resources to do it.

Without input from the CSM, it's unlikely the facility manager would attempt any threat or risk analysis. Just like IT managers, their job is to keep their department or function running. It's rare that people get bonuses for security or add it to their job description on their own.

The CSM can be an advocate for facilities to help them analyse their risks and to advocate for their budget requests. Fortunately, some facilities departments do have large budgets. They maintain brick-and-mortar buildings and assets, renovate buildings and manage the human power necessary to get it all done. Their budgets must be big enough to handle this, so it's often not a big deal to slip some security items in as well. The facilities team just has to know what to include.

The best time to get physical cybersecurity into a building is when it's first built or when it's being renovated. If you're building a $100 million production facility, $100,000 more for physical security isn't that much.

The bottom line is that the CSM needs to get involved with the facilities team, know about any large facility projects and help the company get the security it needs. These are small investments comparatively—and it's money well spent.

CHAPTER THIRTEEN

OPERATIONAL BUSINESS UNITS

So far, we've explored the CSM's relationship to functions that support the business, but not the really big fish—the operational business units (often called BUs). Business units make and deliver the products and services that create the revenue. They make money, provide service, produce value, and run the actual business operations. For the CSM, the continuity of activity for all business units is a primary goal.

The first step toward continuity is to identify the business units themselves. Some BUs might be independent companies within the corporation. When you go to a company website and they list 'Our Services', the services listed there are usually the different business units. Google's business units would include Google Search, Google Ads,

Gmail, GV and so on. Many business units share some support services. If you have ten business units, it makes sense to have one finance department or one HR department that supports them all, though each business unit also has an independent budget.

Operational business units are independent and external to the support units. They are autonomous and have their own leadership. Their leadership teams have a lot of internal power and independence within the organisation. Business units are autonomous in some ways, but they still have a connection to company headquarters. For example, they are usually required to follow the guidelines and policies that apply to everyone else in the company, including attending security training.

A good example of a business unit is a paint factory that covers one geographical area. They make products, sell them and run the service almost autonomously. They have their own budgets, their own leadership and so on. They often acquire some of their own IT systems and their own resources. They usually use company support services like HR to hire people for them that they then manage on their own.

Let's say a hacker was able to infiltrate the main office of the paint company through a secretary's stolen laptop. From there, the hacker gains access to plant management systems and disrupts the entire factory. Or he might disconnect the printers, preventing the factory from producing the paint correctly or labelling the cans. Because of the problem, the lorries cannot leave, there is a production break and continuity of the business is threatened.

Another example is a payment system. A simple denial of service attack that the company is not prepared for can wreak havoc, if the hacker sends a ton of internet traffic to the company's external communication link and overloads its capacity. Then the business cannot process their payment transactions, and people are standing at the counter wondering why their credit card payments aren't processing. These things happen all the time—one retail company, whose main office suffered a DDoS attack, lost the ability to serve customers in 230 stores because their commonly used payment systems could not reach the internet.

Whatever the type of business, continuity risks are the key for production units. It's critical for the CSM to understand the assets that they need to protect, whether it's products or services.

WORKING WITH BUS

The relative independence of BUs means that CSMs must function in an advisory role. The worst thing the CSM can do is try to tell an independent BU what they need to do or that they need to comply.

It's a much better strategy for the CSM to make friends with the director or the head of that business unit and determine what security-related roles are already working inside that unit. The paint factory, for instance, might already have a reliable IT person or an established security person if it's a big enough company. By using that local person as an agent or proxy, the CSM may find it easier to get things done.

The CSM should have a clear channel of communication with the director and head of each of the business units as well, so they can fully understand the security goals and needs of the corporation. It gets complicated. We've seen many situations where the corporation has a common baseline for security requirements for everybody, but they are running many different kinds of business units and businesses. They might make paint in one building and do customer service work in another building. The CSM might wonder if it's even possible to define common security requirements for all the different business units. Are they so different that the heads of some units will say they can't do everything that's required? Quite possibly, yes. The worst response is to insist on the same requirements for everybody, then try to enforce that without flexibility or adaptation.

The CSM can't put security in a box and hand it to the director. Directors have to be involved because they are responsible for their unit's security, not the CSM. If they don't buy into what the CSM is proposing, they could rebel and encourage other units to do the same. When one director says, 'No, security's not important; let's focus on sales instead', their counterparts in other units may follow suit. Now the CSM can get nothing done in any of the business units.

But if the CSM gets the managers involved in their own security, plans can be implemented much more smoothly. Operations managers are goal-oriented, practical people in charge of practical things. If they agree that you need to implement redundant cooling systems, it will happen quite soon. They have their own budget and can decide to implement it. They are bringing in their own revenue stream and have a lot of power, significant budget, and they can accept big numbers. All of this works in the CSM's favour.

There is a flipside, however. If a CSM proposes solutions that don't make sense to the unit managers, the managers might not appreciate it. Let's say the corporate baseline requires running antivirus software on any Windows servers. To most staff, that makes sense, but maybe for production networks running in isolated and segmented production networks away from the company's other networks, it doesn't seem like much of a risk. Management may think that antivirus software is unimportant for them because there's no way for malware to make the jump to their network. They might even have a good point and be justified in refusing to do it.

The decision will depend a great deal on the person and the personality of the manager, which is why getting to know the BU leadership is crucial to the CSM's effectiveness. Any CSM who needs to get something done in a production unit has to be involved with the leadership and encourage them to buy into the plan—preferably with some personal visibility and support.

Cybersecurity managers often fail at this quite badly. We've seen companies with no proper relationship between cybersecurity and the production business units at all. That usually happens when something goes wrong between the CSM and the director in the beginning, and the relationship never recovers.

CSMs should try to get off on the right foot with the BU directors. When starting the conversation, it might be a good idea to leave the requirements at home and not try to force anything right away. Instead, the CSM should ask about what problems they can help with. Then they should shut up and listen and save concerns about fulfilling the requirements for later.

PART III

PROCESS

SECURING EIGHT DOMAINS IN NINETY DAYS

So far in the book, we've talked about the soft skills security managers need—how to interact with people, how to find the right people, how to avoid obvious mistakes. In part III, we'll look at what companies expect the CSM to actually deliver and manage. It will take the CSM through all the obstacles to security with a complete view of the organisation—covering eight domains in an average time of ninety days. These domains—which we'll cover in the next several chapters—are the substance of the CSM's job. Through the domains, we'll explore how to manage security in organisations. Then we'll discuss how to avoid doing everything from scratch.

All the processes and procedures explored in this part

are our recommendations. These are the things the CSM should do, presented inside a framework so that the job makes more sense. The CSM's ultimate goal, whether to secure the organisation or just to comply with a standard, is not important; this framework will work for both goals.

Remember that our recommendations will need to be tailored to your situation; you shouldn't apply the same solutions in different situations. Imagine if a doctor ordered all the medicine on the shelves for every patient— they would sometimes cure some illnesses by luck but would also cause a lot of side effects by prescribing unnecessary medicines. Like patients, every company has to be considered individually. Understand the unique risks at your company first before implementing what we propose. Diagnose first, then prescribe later.

CHAPTER FOURTEEN

DEFINING YOUR POLICIES AND RESPONSIBILITIES

The first of the eight domains we will cover here relates to the paperwork that is necessary for managing cyber-security in the company. Each chapter onwards will provide more details for each type of policy, so read on! Policies might not sound glamorous, but they are the foundation of the security management structure in the company. A CSM who doesn't start here is, frankly, likely to fail. Management support and approval for the CSM lays the foundation for information security throughout the company.

IDENTIFYING POLICIES

In plain English, the word *policy* means a rule governing the way something should be handled. In other languages, the word *policy* can have different meanings, such as the highest level of guidelines or requirements, but generally speaking, people in English-language countries use *policy* for anything that requires a standard of behaviour—a firewall rule, management intent and so on. In a cybersecurity context, the word can also mean any technical, detailed setting that has an effect on security. So when we talk about policies in this book, it's important to understand what level of detail or abstraction we're talking about. Is it about technical policy for a Windows domain controller setting? Or is it management's intent on how security should be addressed in the organisation as a whole? Both of those understandings of *policy* are correct, but they mean different things.

In this book, what we mean when we refer to policy is a written statement, approved by management, on how the organisation or company will react or act in a given area, whether it's cybersecurity, awareness or malware. A policy shows the company's intent and what they strive to do. Policies can be powerful; when management has to sign and approve a policy, they will advocate for it as well.

Policies rarely stand alone; they relate to each other, much like positions in an organisational chart. Policies can be networks or trees of different documents that relate to each other, with branches above and below. The CSM might discover a higher-level policy with management statements and a subject-matter policy below that. The CSM should understand the control structure well enough to see how they relate to each other and if they make sense in that context.

All of the chapters we cover in part III should be covered by written policies, at least at larger companies. Smaller and midsize corporations might rely on one policy that's approved by management at the highest level. Or they might have a security manual or common document that compiles all policies in one place. The bigger the company, the more fractured the structure of these documents usually is.

Every policy should have a handful of chapters; the most important topics include the goal and scope of the policy, what is required and who is responsible. The records must name the person who approved and the date of approval, and may contain version management tables or other bureaucratic details.

GOOD AND BAD POLICIES

Policies can be right or wrong, helpful or harmful, carefully considered or ill-conceived. The CSM needs to know what they are dealing with because an inferior policy can give a bad impression about security. On the other hand, if done properly, policies can boost the security stature of the organisation. Departments usually accept and follow policies that they helped develop and write more than policies that are forced upon them. It's a good idea for the CSM to make a list of existing policies, along with remarks about their quality and acceptance within the organisation. Quite often, the CSM is the one who will be maintaining and developing the policies going forward.

To find out what policies are in place, the CSM must ask about them. When meeting with management, the CSM should ask if there is a policy, if it makes sense and if people follow it. Four possible answers will follow. First, maybe there is no written policy—it's ad hoc. Second, perhaps there is a policy, but no one follows it—it's just a dead paper. Third, the CSM might discover there is a policy, and it's been announced and supported by management, so people know about it. And fourth, even

---

POLICIES VERSUS AGREEMENTS

Policies are not the same as agreements. Policies can be used to direct things inside a company. Policies usually can't be directly used to influence external parties, for example, the supply chain. Agreements are for that purpose.

better, the lucky CSM might find out there is a policy, it makes sense and people are following it.

TYPES OF POLICIES

CSMs, who work with technology and systems, will likely work with, and perhaps create, some technically oriented policies. They also need to be aware of management's perspective; often the main challenge is to translate the will of management into tech speak. For instance, imagine the director says, 'I don't want anyone weird accessing our information without permission. Make it our policy!' The CSM would need to understand that as 'Every user who can access company information needs to have access rights approved by management.' Then the CSM must translate the message further to make it implementable. At that level, the policy says that adding new users requires the IT service desk to request authorisation from direct superiors of the employee and that the user needs to reset his or her password on first login.

Many big companies use Active Directory in Windows environments, for example, as user management for the entire organisation. Security settings, policies and technical policies are controlled centrally from that one centralised system. Since so much depends on it, there might need to be a policy about how to manage security with it.

Similarly, many companies need policies about safely using cloud services. Most companies today are using these services, whether Office 365 or Google enterprise services, or another service where everything is stored in the cloud. In the future, they might not own their email servers or their storage space for documents. When that happens, and it has already happened to many companies, they will need a policy about how to safely use those services and to determine what is acceptable and what is not.

CSMs might be surprised that policies may also be required for third-party services that aren't managed by the company, such as social media accounts. Social media might not pose an obvious security issue, but what if employees act recklessly on Facebook? (Just like Mr. Trump is jeopardising America's reputation with his constant tweets.) That could cause problems for the company, so almost every organisation now has some sort of social media, or 'digital citizenship', guidelines.

Failure to set up policies leaves the company vulnerable. Let's say people use weak passwords when they sign up for online accounts, or they reuse their work password and email address outside of work, like at their fitness club or for online shopping. If there is a breach in the third-party service, hackers will have their work username and password. This happens a lot, along with

Trump-ish behaviour, making these policies necessary. External services come in two varieties: those that the company supports and tries to get users to adopt and those that users choose without company support. We haven't found any company that doesn't have both kinds, and cyber exposure comes from both sources. The company can largely control the first type of service, but the second one is subject to the user's discretion. It only takes a single user—many businesses have had their information stolen because one user decided to store their information on an external file storage and sharing service. When that service was breached, hackers were able to steal the data. It happens a lot.

The company not only needs to make a statement about user-installed third-party services, but a good CSM will try to learn what services people are using without consent or support from the company. Furthermore, the CSM should strive to monitor any external exposure from all external services, supported or unsupported. Otherwise, they will be in the dark about the related risk.

Identifying external services and products is not always easy because many functions once assumed to be internal are now externalised. In the year 2000, for instance, it was common for every company to have their own server room or data centre. Not so much anymore. More than half of those installations have moved to external service providers or the cloud. Some services are now routinely outsourced to vendors, like IT support, front desk reception and help desk services. Many companies don't even own their own laptops any more, but lease them from a supplier instead. All of these services require policies. We predict this trend will continue into the foreseeable future.

SIMPLICITY IS THE CSM'S FRIEND

Given how important policies are, it's understandable that many companies go overboard. We know of a critically important central agency of a government in the Middle East that paid $20 million in consulting fees to create a policy manual and security management structure for the company. In other words, they spent $20 million for paperwork. That's a big overshoot.

When we came in, they asked us to help solve a problem created by all that paperwork—how to implement all those policies in all their complexity. We were a bit cautious in our first few interviews with them. We told them, 'If you want us to help you implement these policies, we will need a copy of each.' We left with a load of documents to read through. When we made a list of them all, it totalled twelve hundred pages. (It was detailed enough to include requirements about how to use the door in the server room and which direction every door should open.) Twelve hundred pages of written policies!

NOT TOO MUCH

Policies are internal compliance requirements. Too many of them will create non-compliance very easily!

Having skimmed through the policies, we went back to the security manager of the organisation and asked, 'How are you going to make sure everybody in your organisation reads, understands and acts on twelve hundred pages of policies?'

It was impossible. We recommended that they throw away between 75 and 90 percent of those policies and leave only the essentials; they could even condense the key policies down to the bare minimum. After all the money they had spent on the documents, they were not happy to hear this. They expected everything to be done by the time policies were ready. We advised this customer to create an awareness programme that condenses the most crucial 1 percent of those policies to learning materials that are relevant to the daily work of people. The rest of the paperwork was fairly meaningless for the staff at large. Money well spent?

The moral of the story is that every organisation needs to find the right level of detail. Having too many policies will overwhelm the employees who are supposed to follow them, and people will just give up trying to abide by them. The policy only needs to include the necessary details.

When the CSM creates policies that dictate things unnecessarily, they will backfire. For example, let's consider a policy that requires that the system administrator goes through all of the login attempts in the company systems weekly and makes a mark after having done it. Because it's a written policy, not one review can be missed. An audit will require a log of every one of those reviews having taken place. If one was missed, even if no harm was done, the auditors will then have to make an audit exception and discuss it in management meetings. CSMs will do themselves a big favour by not creating requirements that may not benefit the company; concentrate on the requirements that matter.

Of course, many policies come from external parties, and there's no way to avoid addressing them. But if the CSM is setting the requirements they themselves have to comply with, it's often better to use general statements in policies rather than specify too much detail in the requirements.

Save the specifics for where they're needed most—areas of real risk. Each policy should be based on some real risk or threat. If there's no real risk or compliance requirements, there's usually no need to implement a policy governing it.

A POLICY BY ANY OTHER NAME?

Policies, procedures, guidelines and standards—what do they all mean?

Not everybody agrees. A lot of companies label things as policies that are actually procedures. Procedures describe the steps necessary to do one process or task. The term should be used when the content is practical. Procedures describe detailed security process steps that must be taken. Quite often, procedures state mandatory things that need to be done.

A guideline, on the other hand, is usually something that *should* be done. It's more of a lightweight obligation. For instance, the company might make a social media guideline that contains two hundred pages of good practices on how to live an ethical life on social media, but it's not mandatory that all employees follow it. They probably should, and if they don't, they may get called out on it, but ignoring a guideline isn't a firing offence, whereas a policy or procedure oversight might be.

Finally, companies, industries and government bodies also set standards—a unified set of requirements for all instances. For example, ISO 27001, part of the ISO/IEC 27000 series of cybersecurity standards, is a standard of good practice for information security that contains requirements that generally apply to all kinds of companies. Though all companies are different, the set of requirements is the same. By the same token, a company might have an internal standard for configuring Windows servers. All of the technical settings could be dictated in the internal standard.

CHAPTER FIFTEEN

CONTROLLING ACCESS

CSMs spend a lot of time thinking about when and how to deny or allow entry to certain systems or resources, from digital access points or physical entryways like IT systems, cloud services, elevators and even doors. Access control is an essential element of security.

Solid security requires control over access to information that matters; the CSM who doesn't take control is not considering what matters and to whom. Anyone who does not have the need to know should not be allowed to have access. Anyone who does need to know should. That's only logical, but it's common to find companies with such strict access control that it prohibits the right users from gaining access, or such lax controls that the risk of data breaches and leaks is increased.

Access control is a way to divide the risk of unauthorised access and data breaches. What people *don't* know will limit what they are capable of doing. Keep your passwords secret from hackers, and you won't be hacked. Keep your business plans secret from the competition, and you have a better chance of winning.

POLICY SUPPORT

Controlling access to information is a complex topic with multiple technologies available. It's necessary to include some good practices and principles in the high-level policy that sets the management tone for controlling access to information. The highest level of requirements could be phased out like below:

> Access to company information is given to people who need that information at their work, but not to others. This is the need-to-know principle. When the need to know ceases to exist, access to information will be removed. Managers in charge will inform the IT department to permit the access to information based on their assessment of need to know.

Obviously, there's a ton of details that could go into more detailed policies for each important system that the company uses.

ACCESS BYPASS

Here's a horror story from real life. Most organisations require some sort of ID, access card or badge to be used in their facilities. Many companies require that employees identify themselves with a lanyard and ID badge that must be worn at all times.

We've worked with some quite unbelievable access control scenarios. In one instance, a school was requiring that all of the parents, teachers, maids and custodians who drop off or pick up their kids wear a lanyard and a photo ID that identifies their face, name and which child they can bring or take out. Nobody without that lanyard should have been able to access the premises and take the kids. The IDs were supposed to be used to control access to the school area and provide a means to check who has authority to take a certain kid along with them. Nice idea, in theory, but practice turned out to be different.

One of their problems was that the school had more than one entry point, usually manned by staff who were supposed to—but in practice didn't always—check the parents' IDs and lanyards.

Every morning and afternoon, the staff came to the gates and greeted parents. They tried to remind the parents to wear a lanyard, but the parents often forgot them at home. The exception was handled by showing the parents to the school office to apply for a day pass. This of course meant that the person could just say, 'Sorry, I forgot', at the gate and be guided to enter the premises. Automatic bypass of access control! At the start of a semester, the staff at the gate was strict about it. Staff asked to see the access lanyards and advised parents to wear them because they're reminded to. But the lanyards were small, and it was hard to see a tiny picture of a face to verify that the person actually matched the ID card, let alone check which parent was matched with a certain kid when they took the kids out. In fact, while exiting the premises with a child, there was no outbound check at all!

As the semester progressed and a few weeks passed, the staff barely glanced at the IDs; they were only registering a colourful lanyard. At the same time, they could not cover every entrance and exit on the premises. Parents soon learned that they didn't need to wear the lanyard or access cards anymore because they could either take a route where there was no staff or just rely on them recognising them by looks. In truth, a parent could just smile and walk past them and say, 'Sorry, I forgot my badge. Bit of a hurry', and nobody cared. That's called social engineering access. Show a friendly face, get people used to it and then enjoy the freedom of access.

Access security lapsed completely once inside the school's perimeter, where the teachers and other staff seldom wore identification. The badges were just for people trying to access. Once they got inside, there was no way to distinguish between internal staff and parents. Nobody could check to see if they were allowed to remain inside because there are too many people wearing lanyards and too many not wearing them. At best, the ID plan gave a false sense of security.

USER MANAGEMENT

The school example illustrates a host of different security problems: overly complex and failing access control scheme, multiple points of entry, lax attitude toward enforcement, lack of formality and training, and no real enforceability. Too often, it's the same with IT security policies.

Access control needs to involve user management, making sure that only authorised users are created and that specific users get access to certain resources and not others. To do that, the CSM has to know who the users and user groups are and what resources are appropriate for them to access. The CSM also needs to know what systems have to be controlled. Only then can they decide what kind of controls to implement. Those are the basic building blocks of access control. Without them, it's impossible to do access control well. Systems that allow centralised control like Microsoft Active Directory are essential in building the access control scheme.

People usually think about access control in terms of horizontal access—how many systems an employee gets access to. The CSM should also consider vertical access and depth of access. Is the user a normal user or an administrator? The more privileged access a person has, the greater their power, and the more likely they are to be a target for cyber criminals, whose goal is to gain broad access rights to internal systems. Do they mention in their LinkedIn profile that they are working for your company as IT administrator? Guess who hackers would prefer as a target?

Most user management today is done with usernames and passwords, and it's clearly inadequate. If we look at the breaches and cyber exposure happening in the world right now, we can see that many of them involve insecure access control processes.

In fact, passwords still pose the biggest threat related to unauthorised access to information, even with all of the security technology available. People use passwords that are too short and too simple. They use the same password for everything inside and outside of the company, and keep the same one for a long time. Recent studies say that half of people use the same password everywhere.

It's not hard to see why people don't use a different password for everything and change it regularly. One individual we met had three-hundred-plus accounts across various internet services. That means three-hundred-plus usernames and passwords. Managing that is, of course, a huge problem. This isn't just a company problem or an individual problem; it's a planetwide problem.

Some companies have offered an apparent solution to password management woes, like using a Google, Facebook or LinkedIn authentication to log in with a click of a button. Users with many accounts appreciate this because they can just use the same passwords for their company and private services whenever they use the internet. This is very convenient for the users but also places a lot of trust on these authentication services and centralises the risk of compromise. If one of these big ones gets hacked, a lot of other systems and information will be at risk. LinkedIn was hacked, and all usernames and passwords were stolen a few years back. Everyone should know, right? But few people know that many of these stolen passwords still work today.

Unfortunately, when people use their work-related email and password—their user account at work—they inadvertently identify where they work, information about the organisation and what services that might apply to. Even if those passwords are encrypted or hashed (in tech language, this means they are protected) in third-party services, once a hacker cracks the passwords (defeats the password protection), he potentially gets access to everything, including the user's work servers that share the same password.

MULTIFACTOR AUTHENTICATION

Login credentials are valuable—they're a sort of currency that can be traded in the underworld economy. Some hackers actually trade login credentials for money or sell access rights to certain companies or types of business systems. Servers and workstations might trade access for some other service. On the dark web, access to a corporate system or to certain servers in a big company can go for fifty-five US dollars.

Companies have tried to create more effective ways to authenticate people—to identify them and make sure they are who they claim to be when they log in to a system. A company might link the password to another factor, like an SMS message sent to the user's phone with a PIN to enter at login. Banks use physical number tokens that generate PINs for you based on time and secret keys. A web bank might issue hardware tokens, or PIN tokens, for its users. A commercial business might be using a virtual private network (VPN) for its users, and VPN software for every user who is working for them. Then they could use a password, username and digital certificate to authenticate the connection and its users.

Layering on additional factors of authentication is usually quite effective; it increases the difficulty of breaching that system. Having said that, there are some instances in which the additional complexity of authentication didn't actually improve security much. Most people are aware of SMS authentication, or the one-time password (OTP) solution, for instance. With OTP, whenever a user logs in to their web bank, they're required to answer with a PIN number that's sent to their cell phone, in addition to their username and password. A hacker using another phone to try to access the account, even if he knows the username and password, can't log in without first getting an SMS, reading the number and using that to log in.

Additional authentication steps like SMS tokens add complexity to the authentication process, and remember, complexity is the worst enemy of security. Even multifactor authentication like SMS tokens aren't foolproof. Anybody working in the company that provides the cell phone connection could intercept the user's SMS. Or, if manual processes aren't strict, a hacker could portray himself as someone else, then manage to open up a clone SIM that receives the same SMS messages. Ironically, that supposedly super secure multifactor authentication scheme that combines an SMS token with a good password could be compromised by the same feature that's meant to protect the user. How many little shops are working for your mobile service provider and are able to issue cloned SIM cards or change the ownership of the mobile connection? Try enforcing an access control policy on them!

Complexity rarely makes security better. That's not to say that adding a second factor of authentication is a bad thing. It's good, but there are limitations. There are a lot of different ways to authenticate people. The lesson here is that the more secure the system needs to be—like a bank, for example—the more security is needed in authentication.

USEFUL TOOLS

Anyone who has a lot of passwords and usernames—say, more than ten—should get a password management tool. They are usually referred to as 'password wallets'. There are several available that can run on a laptop and sync with a mobile phone. For a few dollars a year, all of a person's passwords can be securely stored in one wallet, so they just have to remember one master password for the wallet. Then it helps them log in and authenticate to different solutions and stores them securely. Having a password wallet makes life easier for users.

At the same time, we have to remind users that wallets also pose a risk, especially online wallets. Passwords are stored in an encrypted file, and in some solutions, that file is sent to a central repository on the internet. If that application or an individual's computer is hacked, all of those passwords may be compromised in one place. Still, online wallets are an effective solution for people who have a lot of passwords and companies that have a lot of users, even though it has a single point of failure. Anyone choosing an online wallet should choose one that's been thoroughly tested—by more than one person or one company. It needs to be more than just a convenient solution. It needs to be a secure one.

We couldn't begin to cover all of the options for access control today. There's a multitude of authentication and access control technologies, services and solutions that go under this topic, and all of them would solve bits and pieces of the whole problem. No single solution will cover all of the access control needs of the organisation. This is because no single service can be compatible and integrate with all the various services out there. The CSM's job is to gain understanding of which access control technologies are a good fit for his business and to help IT to design a scheme that is flexible, has good coverage and is able to secure the business well enough.

AVOIDING SECURITY THEATRE

For staff to use passwords effectively, they will need to understand what matters in password management.

WHAT'S SO SPECIAL ABOUT THOSE SPECIAL CHARACTERS?

Why did we start using special characters in passwords in the first place?

The argument for them comes from mathematics. In theory, adding special characters increases the workload needed to defeat password protections (encryption, hashing and so on).

The thing is, we don't care about the math debates about password complexity. We care about the outcomes: if it's possible to be a happy password user while making it impossible to crack the password.

Instead of making passwords hard to remember and enter, just make them loooooooong but easy to type and remember. 'Oh dear a black swan crossed the road' is actually a very good password. Password length is the single most effective way to make the passwords secure. Remember, hackers have no way of knowing if you used special characters, uppercase letters, numbers and so on in your password. They have to assume that all types of characters were used when they try to crack yours. Besides, a password over twenty characters long is virtually undefeatable by any practical means.

There's a lot of conflicting information out there. Many government and corporate guidelines, for instance, say that a password has to be eight characters long. It has to contain a mix of letters, numbers, uppercase and lower case and a special character, and it has to be changed every thirty to ninety days. However, research and practical experience has shown that there's one property above others that makes passwords strong: the length.

Some argue that the complexity of the character set is also significant, but it's not as effective as sheer length. A long password is a strong password. Researching this subject will reveal a lot of academic papers and calculations pointing to different directions. But hackers think differently; they are only interested in defeating the password protection by any means necessary. From an attacker's perspective, only that outcome matters, not the computational difficulty!

Even poor password policies can seem fine on the surface. That's a problem because people will think their login is secure when it's not. If people think they are safe, they will drop their defences. They figure, 'We already did this two-factor thing. Nothing else can hit us.' Part of access control is spreading the best practices of security and managing the sense of security. Or maybe they think that their Windows AD policy requiring ten-character passwords with all the complexity is good enough. Guess what? It isn't! If a hacker could crack it, it's no good, and that's the only metric that matters.

What's the security theatre then? Ineffective password policy is like airport security, when air travellers have to take certain items out of their bags, like the liquids. It's not because they need to be scanned separately but because security wants people to participate in the security process. When you participate, you feel like it's effective. This is called 'security theatre'. It's the same with two-factor authentication, bad password policies and so on. Users who have to type something extra feel like they're actually part of the security process. Unfortunately, two-factor authentication won't prevent someone from listening in on mobile phone calls or tracking where people internet surf. People just think it's safe because they're participating in the security protocol, while in reality, the threat still exists, and the security can be defeated.

Companies should make efforts to train their people and to enforce proper security policies and procedures as much as possible. Enforcement usually means setting technical limitations and requirements for passwords, but not the ineffective eight-character codes we talked about earlier, with or without special characters. As we saw earlier, it's the length of the password that makes the biggest difference. Now go back to www.sha1-online.com and try something like 'my password is very secure', and Google it. No findings, right?

If users are, for example, advised to make at least twenty-character-long passwords, using a poem and some sort of a string to add to that poem, like a system name or something only they can guess, passwords will become so impenetrable there is no way to crack them, even if they leak.

IS IT CRACKED ALREADY?

Try this easy trick yourself. Go to www.sha1-online.com and type in any password. You'll see a long string of characters as a response. Copy and paste that string to Google. Did you get any search results? If you did, that password has been already cracked somewhere out there! Here's an example. Try this password: P@ssw0rd—you'll get this response: 21bd12dc183f740ee76f27b78eb-39c8ad972a757. After Googling it, you'll see that there are many results. This password would meet most of the password complexity requirements but is very unsecure.

SHA-1 is an outdated algorithm, and using it to secure passwords is a really bad idea. Yet, a few years back, LinkedIn used this method for their password security. And when the service got breached, all the passwords were easy prey for hackers.

The bottom line is that you can't know if your internet services are using good password protection mechanisms or not. But you can use a good password that can't be cracked even if someone is dumb enough to store it with something silly like SHA-1.

When the company sets a password policy, it affects users' behaviour not only in the office, but in their personal lives as well. If the company says, 'Eight is enough', people will use eight in their personal lives. If the company mandates twenty characters, maybe their personal passwords will also become longer. That's important because security exposure also comes from the employees' and management's personal lives. Make 'twenty-plus' their mantra.

Long passwords sound like a pain, but they are actually

easier to remember; the user can type something that makes sense to them. It doesn't have to be random. It could be as simple as what you usually buy from grocery stores: 'my favourite milk is from Australian cows'.

One tip: if a user keeps the same recipe for their shopping, they can just add the system name. If it gets breached, no one can use it because it won't match with any other system, but it will still be easy to remember. Example: 'google.com is my favourite milk'.

CRACKING THE PASSWORDS

There are several approaches to cracking a password, including dictionary attacks, rainbow tables and brute force methods.

DICTIONARY ATTACKS

A dictionary attack uses all the words in all the languages in the world, as well as millions of leaked passwords from data breaches. When a hacker has obtained an encrypted form of a user's password, all he has to do is to take that long dictionary and hash the words in the dictionary. (Remember that www.sha1-online.com example earlier? Same idea but just faster!) If he finds a match, he knows that this was the clear-text, human-readable password of the user. A variation of this technique is when the computer sifts through all of the dictionary words and tries the words with tiny changes, like an exclamation point or a hashtag or something linked to the words. A normal computer can make millions of guesses per second, and cloud services can do it in parallel many times faster.

BRUTE FORCE

The next method is brute force. Here, the hacker uses as much computing power as they have, then they start blindly searching all possible existing passwords, perhaps like A, AA, AAA, AAAA and so on, and with different lengths of the search, doing millions of guesses per second. The idea is to try until the produced hash value is the same that hackers stole from the victim. Then they would know that it's the same password!

The brute force method takes a lot more computing power and time, of course, because it requires going through all the different possible versions of passwords. That's where the name comes from—it requires a lot of brute force! Hackers do this when the easy way doesn't work. They use stolen credit cards to buy Amazon accounts, then use cloud computing servers to crunch the numbers and try to crunch as many passwords as possible.

The longer the password, the harder it becomes to crack it by brute force. A password of 'ilikegoingtothebeachonsaturdays' works because going through all the passwords that long will take literally forever using a brute force method. But if you use short ones, with eight characters or alike, no matter how complex they are, given enough time, they will be cracked. And sometimes brute-forcing is just work that can be skipped entirely. Enter rainbow tables!

RAINBOW TABLES

Another type of password-cracking method puts the dictionary attack on steroids—it's called the rainbow table. A rainbow table is a pre-computed version of all possible passwords that can exist up to a certain length. A rainbow table would start with short and simple passwords like A, B, AA, AB and so on, and contain the corresponding hash values of these passwords. These tables are huge, usually terabytes in size. They are powerful because a hacker can do one lookup to his table, find the corresponding password hash value and see directly the corresponding human-readable password. This method compromises passwords quickly—in a fraction of a second. And the table only has to be created once, though it takes terabytes to store. Typically, a rainbow table would contain all passwords up to a certain length, something in the order of eight to ten characters long. And because everything is nowadays cheap in the cloud, a hacker could just go and search all existing rainbow tables online, as a service,

without bothering to store or create the tables himself!
This is the final nail in the coffin of short passwords, no
matter how complex they may be.

FAMOUS BREACH

We read about hundreds of major breaches in the news
every year. In 2012, for example, around 170 million
LinkedIn user accounts were breached. The accounts of
170-million-plus people were available to hackers. The
majority of these passwords were easily cracked in no
time by using rainbow tables and brute force techniques.
Soon, lists of cracked LinkedIn passwords started circu-
lating around the dark web. Many of these users were not
aware they were compromised and did not change their
password, or chose to keep the old one, perhaps not in
LinkedIn but in other internet services they used. Now
hackers had access to a multitude of these accounts and
passwords at LinkedIn and many other internet services
where users were using the same credentials.

On the surface, it seems that this is solely a personal problem for LinkedIn users, but in reality, it came back to bite
a lot of businesses too. Those compromised accounts
were used to collect personal user information, create
fake messages to lure people into clicking phishing links
and other kinds of fraud. Success rates of these kinds of
attacks were fairly high, as hackers were basically exploit-

ing trust that people place on each other's social media profiles. When messages are coming in from a user's real LinkedIn or other social media profiles, and he sends you a link, will you be induced to click it? Of course!

Even worse, although LinkedIn is a huge company, it used a lousy password-encryption technology back then, just a plain and simple SHA-1; the passwords were not protected against rainbow tables, brute force, or dictionary attacks, although this should have been a very basic thing to do for any security-aware software developer.

Then, since people were using their business credentials to log in to other systems, hackers were able to use them to log in to many of those business systems as well. That massive external LinkedIn breach led to a multitude of other breaches. It was, and still is, like one big avalanche that never ends, moving from one service and victim to another.

At the time, this breach was titled the worst breach of the decade, or even throughout history, because the exposure was so huge, and the quality of stolen data was high. LinkedIn is not an isolated case either. A normal week in cyber intelligence services starts when we see another few hundred million accounts exposed in one internet service or another. This unfortunate trend isn't going to get any better anytime soon.

CONSEQUENCES OF OVERSIGHTS

Let's look closely at access control security issues in a company we worked with. This company had thousands of people set up on a Windows network. We worked with the twenty-member IT staff, each of whom had access to some of the servers in the network. That level of access was appropriate; many of these people needed administrator or root-level access almost daily in their work. A few of them had the highest-level privileges and access rights to every system in the company, and that was justifiable because it was their job. Naturally there were times when more than one person needed to access the systems, so they had to share some administrative passwords between the team members.

With twenty people and two hundred servers, there were a lot of usernames and passwords to remember. The complicating factor was that all the accounts in question were prime administrative accounts for all of the systems in the company. They had a username for each system, then different passwords for different users, and so on. Suddenly, they had the same problem we talked about with individual user accounts; they needed a solution that would allow them to share the passwords and store them somewhere. We've already mentioned a password wallet solution earlier, but this wasn't their answer, unfortunately.

The company's solution? They set up a Windows shared

folder in the network, which is a folder that users in the same network can open on their computers. The shared folder was accessible to the IT team members, and they could all edit the same files in it. So IT put all of their passwords for those two hundred systems and twenty IT professionals in one Excel file in that shared folder—one file containing all of the passwords in the network.

The new system was convenient from a usability perspective, but the company overlooked the risk and impact this setup caused. One beautiful day, someone in IT forgot that he made that folder, and all files inside were shared to all users inside the same Windows domain. They could have limited access to that shared folder, where only those twenty people could open the file and use it. They could have also encrypted the file so that only people with special decryption software—like a password wallet on their computer—could have opened it. Even if a hacker had gained access to it, they couldn't have decrypted the file.

Instead, anybody in the company could log in to that folder, open it, open the file and look up the main user account, password and username for any system in the company.

So what happened? Hackers penetrated the network—they used the LinkedIn breach to log in to one account belonging to a C-level executive. From that account, they fabricated a phishing message and sent it over to a few select individuals in the company. These people were naturally inclined to click the link and got their computers infected by remote-access software. Now the hackers had access to the network. The first thing they did was crack the local passwords of these users. Next, they proceeded to look around inside the network, looking for anything interesting or of value. After these initial steps, they used something called Windows PowerShell to automatically look for network folders and systems within the network. Kind of like mapping the terrain where they found themselves. Of course, they found the shared folder, named conveniently 'IT passwords', and the Excel file where all of the passwords were located. This led to the compromise of all two hundred servers. Now all they needed to do was to log in to all those servers and install covert remote access programs called rootkits on each of them. Now that they had even better access to the servers than the administrators, they simply exfiltrated all interesting data from the company systems.

The company learned about this incident the hard way—someone contacted them for ransom, asking for money, otherwise the hackers would leak all of the information they had stolen.

Otherwise professional people failed to control this problem. It was a huge disaster at the time, and it took a lot of work to clean it up. A rootkit is so stealthy that you cannot know if it's still there. They had to spend a lot of time and money to fix the issue, blocking communications in and out, reinstalling a lot of their servers, changing passwords and so on.

CHAPTER SIXTEEN

RISK MANAGEMENT

Risk management is a practice, a profession and a bit of a science. It also has measurable, real-life effects.

WHAT IS RISK?

The basic formula of risk says that it is the probability that something unwanted will happen multiplied by its total impacts. In a personal context, it means if I cross the street, there's a likelihood that somebody is going to hit me with a car unless I'm careful. So I'm making a personal risk-management decision by looking both directions before I step off the curb.

In cybersecurity, risk is often approached from a technical angle. We can measure risk by assessing exposure, vulnerabilities, likelihoods, exploitability, impacts and so on. For example, how many people do we employ that could be fooled by phishing attacks? Or how many systems do we have online that could be hacked? Using that information, we can then do technical measurements to deduce how high our total IT or cyber risk is.

Risk = Probability × Loss

Risk management strives to give quantitative value to risk, but it often requires significant qualitative analysis as well. Even if the risks are assigned numerical values, each one is subject to the opinions of a variety of people from different departments. Each will likely give a different answer.

If the risk managers try to use one single method to assess all types of risks, the formula won't be applicable to all situations and won't create meaningful results for the company. In fact, a professor we know did a review of different risk analysis methods to find out how many different types he could identify. He found more than one hundred different formulas for calculating risk. So, in practice, a CSM must deal with an indefinite variety of risks and dozens of possible ways to analyse them. Remember, risk management is not just quantitative but qualitative too. The CSM must always consider additional ways to describe the situation because reducing the problem to a single number filters out important information.

ASSESS RISKS REGULARLY

Risk management is most effective when it's done frequently and continually, with the lightest possible footprint, then updated and revised on an ongoing basis. Too many companies do risk analysis once a year, update it for the following year, then forget about it until the next year comes around. That doesn't work.

Some periodical reviews of risks are necessary, though. We've found that it's quite effective to meet quarterly with the key people responsible for operations, leadership and management. In these quarterly meetings, the participants should spend fifteen minutes to half an hour talking about the most obvious major risks, then write down the results of that conversation. Regular communication keeps the CSM in the loop—management is often aware of existing and emerging risks that might not be obvious to the CSM.

On the other hand, leadership won't know about all the risks in every department. If the CSM is working with a company with no other risk management people on board, it's his or her job to ask the different departments to do their own risk analysis work periodically and then report their findings to the CSM. (If the company already has other risk management people, the CSM would be wise to use their services instead of trying to run the whole show alone.) When all known risks are pulled together, and the largest risks are communicated up the command chain, this is called risk consolidation. Smaller risks are left down to be handled at the lower branches of the organisation, and bigger ones are communicated upward. The CSM should understand this process and know how it works.

RISK MANAGEMENT PROCESS

Risk management is a process, and we usually start by identifying and assessing risks. In a smaller company, the CSM might arrange a workshop all by himself. In that workshop, people come together to identify potential risks. The workshop should start with identification, then proceed to discussing how likely and harmful the risks are.

After that, we decide what to do about each risk—mitigate it, avoid it, share it or even accept it—we call these risk management options. Mitigation means making risks smaller, for example, by implementing an additional security technology or a process so that the risk becomes either less likely or its impacts are alleviated, but the risk

MOST COMMON RISK MANAGEMENT OPTIONS

1. Avoidance (stop doing the risky thing in business)

2. Mitigation or reduction (optimise–mitigate)

3. Sharing (transfer, outsource or insure)

4. Acceptance or retention (decide to keep it)

may not go away entirely. Avoidance means rejecting the risk altogether by pursuing that avenue. If we know a country is at war, we choose not to travel there. That's risk avoidance. Risk can also be shared between parties, for example, the company and its insurance company—for a price. Or senior management may want to simply accept a risk and sign for it. This is also an option!

UNIDENTIFIED RISKS

One of the biggest threats to security is management that doesn't even realise there is a risk. Or, if they do recognise a risk, they don't take ownership and assume someone else will handle it. For example, in a factory, the production manager may not feel that business interruption risk is their problem if data connectivity goes down. He believes that's an IT problem, so he doesn't bother to build redundancy. When the data connection line goes down, he blames IT. He may avoid blame, but he doesn't avoid the threat itself; the production line still goes down.

Many managers, even if they recognise the risk and agree who owns it, may not comprehend the magnitude or possible impact of the risk. Too often, risk is underestimated. For example, it may be fairly simple to estimate the probability of a communications blackout, since it might happen once or twice a year, maybe three times. If it happens that often, it's quite probable it will happen again.

What is the impact of these blackouts? By collecting the data on the previous incidents, a CSM can answer many questions about impact: How much did it cost to fix? How much damage was caused? Was it direct or indirect damage? Answering those questions provides a clear picture of how much it could cost in internal labour, incident management, external costs and so on. The nice thing is that when you can quantify a risk, you can ask for a budget to address it.

QUANTIFIABLE VERSUS QUALITATIVE APPROACHES

Risks like damage to reputation or hindering future growth are hard to quantify, and yet these subjects are top concerns of business owners. Successful risk managers or CSMs have to address both. If they cannot put a figure on it, they'll have to sell their plan another way, like telling the story of how things are likely to turn out if the risk is not properly addressed and then defending the estimation of how serious the risk is.

For new business ventures, like a startup that's launching a new product line, it can be useful to ask them to identify scenarios that could be catastrophic to the business. Help them think about what could be bad enough to bring the company to a standstill. Scenario analysis like this will drive your point home. The simplest kind of scenario analysis is when people come together in a workshop and come up with causes and impacts in a catastrophic business risk scenario. Let's say they are setting a joint venture with a business partner. In the workshop, they could identify indicators for failure of that venture. If these indicators should appear later, it might mean that the scenario is materialising. This is very close to business management, and a purely qualitative approach to managing risks.

HOW ABOUT QUANTIFIABLE RISKS?

The ultimate goal of risk management is to quantify risk and make it relatable to the business. Risk management is about minimising the downside of business decisions on all levels of the corporation. That allows the company to take more risks and to take advantage of more opportunities.

Risk managers ask questions like 'Can you tell me how often this potential risk turns into an actual problem? What's the probability? What's the impact in terms of money, life, property, downtime or business inter-ruption?' They will be interested in qualitative and quantitative data.

Risk management people will first identify risks, then try to understand them, assessing each risk either quantita-tively (in terms of money) or qualitatively, if the numbers are vague. They will put the risks in order of priority, with the bigger ones at the top. The CSM should meet risk management people and make them explain how risk management works in the company, including where to report risks, how they are assessed, and how to partici-pate in the process.

EXAMPLE OF A QUANTIFIED RISK

For example, if the business has been experiencing network outages, this is clearly a risk that needs to be assessed and explained. Five outages per year, causing $100,000 loss each, means a $500,000 per year cost.

Investment-wise, it would make perfect sense to put money into network resiliency and enjoy a good return on the investment going forward.

PUT THE RISKS INTO THE RIGHT BUCKETS

When the CSM is faced with prioritising risks, it can be helpful to group risks by category. There are many ways to categorise risks. We're using one traditional four-category system here, but you might want to look up other frameworks as well, like COSO ERM.

The first category, and the most obvious one, is hazard risk. That means things like a fire, flood, tornado or heavy rain. These risks are usually well managed and covered by regulations. For example, automatic fire extinguisher systems are required in office buildings. There are a lot of hazard risks to control when setting up a server room!

The second category is operational risk. This category includes things that companies do to limit their exposure to operational risks, like communications breakdowns or IT system breaches. Locks on the door are for operational risks of burglary in the office. You're not really required to lock the doors by law, but you still do. Many of our commonplace cybersecurity risks and related technologies fall under this category.

The third category is financial risk. There are many different types of financial risks, including theft, counterparty risk, foreign exchange risk, market risk, cost of capital risk and so on.

The fourth category is strategic risk, which encompasses larger issues, like growth strategy risk, branding or image risk, competition risk, customer and industry strength, change risk and the risk of becoming obsolete. One example of strategic risk is a technology risk. When it materialises, the company's products may become obsolete in a short time because of leaps in competing products. Nokia was a victim of technology risk. In a few years' time, they disappeared off the map.

RISK REGISTERS AND GOVERNANCE

One of the tools used in risk management are risk registers. It might be called something else; for example, it might be part of a GRC framework (governance, risk and compliance). GRC is usually managed with a specialised software, but the concept of a risk register is the same: it's covering a list of risks, among other things. A team of analysts maintains the risk register, maybe within an application or an Excel spreadsheet.

The concept of a risk register is sound, but if the risk management function isn't focused on one specific type of risk, the team will collect a wide range of different types of risks in that register, and it will become unwieldy. Risk management for the sake of itself isn't very useful. For instance, say one department has an antivirus problem, a different department has a financial fraud problem, someone else has a health hazard, while a VP is talking about business venture risk and strategy initiatives. All of these risks, each with different impacts and probabilities, will end up in the same risk register. The register will then become bloated. It turns into the final resting place of a collection of bad things in the company that nobody is doing anything about, rather than a call to action. One case we have seen involved a financial institution that used their risk register to collect everything in one place. The register had fifty columns to be filled out for each risk. These people missed the point of why risk registers exist. They exist so that risks can be compared and communicated effectively, not because risks can be managed through the tool. A risk register is not a management tool. In fact, it's very difficult to even manage a list of that size; it would require a full-time position just to maintain that spreadsheet.

It's important that the register doesn't become a dumpster for all risk items, because it's too easy for management to forget items once they're relegated to the register. The risk register shouldn't become a bloated monster. If it's too big, there's no way to bring it to the table for people to talk about it. In reality, there's always more risk than you can actually handle, but if there are too many risks on the register, they can't be prioritised effectively. Instead, select only the top handful that you want to act on and focus on those.

VISUALISE WITH A RISK MATRIX

People in risk management often use a risk matrix to estimate and rank each risk in relation to other risks and then prioritise accordingly. This allows them to visualise different risks and present them more understandably to senior management. A risk matrix can give a hint to which risks should be handled first. There is a pitfall though. A five-step risk rating system like the one below looks good on paper. But how skilful are people really in assessing the likelihoods of different risks? How about their impacts? Give the same risk analysis task to several people independently and you are sure to get different answers.

	IMPACT →				
	NEGLIGIBLE	**MINOR**	**MODERATE**	**SIGNIFICANT**	**SEVERE**
VERY LIKELY	Low Med	Medium	Med Hi	High	High
LIKELY	Low	Low Med	Medium	Med Hi	High
POSSIBLE	Low	Low Med	Medium	Med Hi	Med Hi
UNLIKELY	Low	Low Med	Low Med	Medium	Med Hi
VERY UNLIKELY	Low	Low	Low Med	Medium	Medium

(LIKELIHOOD ↑)

RISK PRIORITISATION AND PLANNING

If there are twenty risks listed, the CSM might focus on three of them in the coming year. When they get management approval to pay for fixing all three of them, they've still got seventeen left for next year. The CSM must make sure the owners of those remaining seventeen risks understand and accept this. The CSM needs to say, 'Hey, we have all these risks that we've decided not to do anything about this year. Let's handle those in the coming years. Agreed?' After that moment, the CSM is not responsible for those risks or their consequences should they materialise. We call this technique residual risk approval. Without residual risk approval, if the CSM doesn't notify anyone about a risk of ransomware that's on the register but not in the top three, someone might later blame the CSM for not telling anyone about it.

> Management approval of residual risks lets you off the liability hook.

In essence, the CSM must do three things: collect the right information, give it to the right people and do it at the right time. Senior management expects the CSM to keep track of all the risks, alert them of which ones are most important and do it at a time when it can be budgeted for and acted upon. If you can do that, you will succeed as a CSM.

Another challenge is that people tend to exaggerate the negative effects of risks, at least until it comes time to pay for fixing them. Then suddenly it doesn't seem critical at all. So which is it? The CSM has to sort through that confusion to get down to the few risks that actually are critical.

The CSM's first job is to find the critical systems—the crown jewels of the company. If these systems shut down, the business would shut down. They should be easy enough to identify, but as the CSM asks people questions like 'How critical is your system to the business? What happens if it's not available for a few hours? How about a day or two?' the responses may all sound alike. All of the systems are considered critical by the departments that use them. But looking from the perspective of the whole enterprise, they might not be critical at all!

It's up to the CSM to assess the risks professionally and without emotion, and to help management prioritise them based on facts, not beliefs or opinions.

CHOOSE THE RIGHT METHODS

With hundreds of formulas for calculating risk out there, we have to be able to quantify and assess them, then implement the most appropriate one for our needs. As previously mentioned in this book, generally speaking, a risk formula takes the probability (P) and multiplies it by impacts (I), and that will equal risk. It's not very complicated, but it's only a starting point, and it is still how many companies do it.

There are hundreds of different formulas or more. The CSM has to look up the few that are applicable to cybersecurity. He should understand how they work, why they are used, where they are used, and when not to use them. Then he can figure out which ones are applicable to which situations.

If you take a risk formula from an insurance company and use it in a manufacturing facility, it won't work. It's important that the right risk assessment methods and formulas are used for the right targets. For IT, use IT methods. For financial risk, use financial formulas for risk. If there's a fire risk, you would use formulas applicable to fire risk management.

Ultimately, once you learn how to use risk management methods the right way, you can manage the downside of any opportunity and make it more likely to succeed and to turn out to be profitable.

Studies have done quantitative analyses of different companies with very large sample sizes. They found out that the companies that do good risk management average a few percent more profit than the ones that don't. A few percent greater profit is a huge deal.

CHAPTER SEVENTEEN

COMPLIANCE AND ASSURANCE

By now, our readers know that compliance does not equal security. It's not necessarily good for the business; not always, anyway. Everybody in the industry has to do it, but doing it too diligently can actually reduce your competitive advantage. We think of compliance as a necessary evil.

The big question for a CSM is how much compliance is mandatory and how much is voluntary. Compliance can mean meeting the requirements of the law or exceeding them. It can get confusing; legal contracts, for instance, may be subject to certain rigid requirements and also to some requirements that can be negotiated or renegotiated. The CSM has to know which is which.

COMPLIANCE VERSUS ASSURANCE

Compliance involves finding out what the minimum requirements are and deciding how to meet those requirements. Assurance, on the other hand, is making sure that compliance requirements have been met. Let's say some government department wants to audit your compliance with their requirements. Assurance would be gained when they send an auditor over to check up if things are in order.

Compliance is binary—either you're compliant or you're not. Even if you're 99.9 percent done, you're still partly non-compliant. If that vague line of requirement hasn't been fully met, even though everything else has, you're still not compliant.

The way compliance with requirements is verified will depend on each standard. Some actions are mandatory in all instances, while others are negotiable. Auditing is required to validate compliance.

Once everything is audited, validated, and perhaps even certified, the company can demonstrate to its customers and partners that they are compliant. If a company can do that, it's easy for the customer to consider the company safe, and other companies will feel comfortable doing business with them.

WHY COMPLIANCE FAILS

When compliance fails, it's usually for one of two reasons. First, many companies go too far—way above and beyond the compliance standards—and it ends up costing too much money. For example, if the compliance project planning is done by security people, they would want to cover everything that would affect security, whether it's required or not. Letting specialised departments handle compliance planning is a major mistake. Compliance should be overseen by a project manager or director with a financial or business focus. Don't overdo it, and don't let people mix their personal agendas with it.

> A good practice is to have a separate budget for compliance projects and a dedicated lead for it.

The other reason compliance fails stems from misunderstanding the implications of compliance contracts. Too often, companies sign a contract without realising what every clause in the contract means. There might be additional compliance standards named in the contract; once the agreement is signed, the company is now required to meet those standards in order to stay compliant with the contract. Unnecessary requirements often creep into an organisation this way.

REALISTIC TIMING

Compliance takes more time than companies often realise, not necessarily because the standards are hard to meet but because companies don't know how to deal with auditors to the best effect.

If the CSM can advocate to get the auditor involved early in the process, things progress much faster. Calling them in early on, showing them what the company is going to do and giving them whatever documentation they need as soon as possible will speed up the process.

ONE SIGNATURE TO RULE THEM ALL

Companies should approach compliance with the mindset of doing what has to be done with optimised effort. Understand what's required, implement it efficiently, don't do anything excessive and get management to sign off on it. That's it.

MAKE IT EASY FOR AUDITORS

A note on compliance documentation: it's best to design it so that it advances only those questions pertaining to compliance requirements and nothing else. Have your documentation follow the same structure as the compliance standard does. This way it's easy for the auditors to find answers for their requirements.

That last part can be a challenge, but a lot of compliance documentation must get approved by top management, in writing, so there is documentation at the ready when the auditor asks to see it. The CSM will bring the written documentation to the appropriate level of management and ask them to approve it, again in writing.

What if there are twenty different documents that need to be signed off by different positions? Does the CSM really need to go to twenty different places to collect twenty different signatures? Not necessarily; that takes too long and isn't practical.

Instead of bugging leadership over and over again, one signature can often cover as much as 30 to 40 percent of all compliance requirements related to security management!

THE MAGIC MANAGEMENT MEMO

The best way to achieve compliance with a lot of the requirements is very simple: take a standard and make a list of all its requirements. Start marking down the things that need to be accepted by the management, and create a management memo with a list of those things. If you have to get policies approved or reviewed, they go in. If you need to have things that aren't compliant, put them on your risk analysis sheet. Yes! Put everything in there that's not actually compliant and explain what kind of a risk exists if each item on the requirements is not done. This is how you would have management approval of residual risks, or things that could prevent you from being compliant. You can work with the management assistant or secretary to make that memo and set up a meeting with the management. Then send over the memo with all the attachments to be approved. Once at the meeting, you need to explain to the management that when this memo is signed, all related documents and residual risks will be approved at once, and the company will get many compliance requirements done with a single signature. Explaining this in person is very important so that the management understands that getting this memo approved is a matter of gaining compliance, not a matter of lengthy debates. It's either going to be approved and we're compliant, or they can decide not to be compliant. Easy call.

We have been able to pass difficult audits by very strict bodies like defence forces in as little as ninety days by using this simple approach. We gave the title to this part of the book because we know it can be done, and we have done it. Our results with our clients have been extremely good. Auditors like that the documentation, risks and all the approvals are presented in an orderly fashion; everything is easily related to their requirements; and any exceptions are communicated from a risk perspective to them. And everything has management support in writing. Quite often, the auditors have made comments like 'This was the fastest audit and compliance project we've ever seen, with the best results among all the companies we've ever audited.'

CHAPTER EIGHTEEN

BUSINESS CONTINUITY MANAGEMENT

In an ideal world, the IT cybersecurity manager will know the major continuity risks a company faces and will have preparations in place to make sure nothing that disrupts critical processes ever happens. But if a disaster does occur, the CSM needs a plan for that as well. We're covering business continuity management in this chapter from a cybersecurity angle. The CSM won't be alone and should cooperate with other business units to cover other types of continuity risks besides cybersecurity.

Planning for those disasters is also known as business continuity management, the high-level planning for how to recover, restore normalcy and minimise the loss after a serious incident. For instance, if you're running a factory and there's a flood, you might have to pause production for a week. That's major. Water damage might be one risk; fire might be another. Of course, these types of disaster are improbable, depending on where your business is located.

Most companies, however, face the risk of an IT problem disrupting their factories or core processes. Most businesses depend on IT running smoothly. That means computers, chips, software, networks—even the flow of money depends on everything working in concert. The question is, when something major disrupts IT and, in turn, interrupts critical production and processes, how will you react, minimise the loss and restore operations to normal? And what preparations can be made to make sure nothing like that happens?

The simplest scenario that affects just about every business would be a communications breakdown. As we know, almost every company is using the cloud. This means that if a factory is up and running, it's connected to the cloud, and its business processes depend on the cloud. What happens when there's no network? It can't help but disrupt some part of the business.

Business continuity management requires looking at a number of different pieces: the network connecting the offices, the production sites and the data centre or cloud are the obvious ones, but not the only ones. Another curious area to focus on is the domain name system (DNS). This is usually critical to everything happening inside networks. If the DNS fails or someone takes it down, the systems cannot resolve host names. They cannot find each other unless they are using an IP address for connecting. Similar systemic dependencies exist in most IT environments. The key is to identify single points of failure and build redundancy to counter scenarios that could bring the business down to its knees.

In a perfect world, the CSM would have a budget to cover double and triple backups or to run a redundant server for all systems. Few do. CSMs have to at least cover the critical functions; the difficulty lies in knowing which parts are critical.

TESTING IS ESSENTIAL

IT continuity management requires preparation, planning, testing, practising and updating. A lot of companies focus on the plans but neglect the testing and practising parts because they're harder and more expensive. But without testing, the company has no assurance that the backup plan actually works. Without practice, people won't have the skills and experience to do what's necessary to get everything back online in the event of a disaster.

It's not unusual for a company to sink $1 million into redundancy and backup hardware, or even establish a secondary disaster recovery site for IT, without ever testing the system. Will it work if something happens? No one knows.

We've seen numerous examples where a company had a primary processing site and a secondary site—a hot site and a cold site—and they never tried to switch over and operate it from the other side. If they don't even know if it works, what is the point?

LEVELS OF PREPAREDNESS

Disaster recovery preparedness exists on a spectrum; at the lowest level, the company does not have any redundancy. Everything's running on one site, and that's it. If a system or network goes down, they're pretty much sunk and forced to recover any way they can. Quite often, this level of preparedness makes recovery slower; perhaps it takes a few days to get everything back on track.

The second level of preparation is to have some level of replication in place, like a cold site that holds data from the systems in a separate physical location or even in the cloud. Then, if there is an event, the other system can be brought to life, and network traffic can be routed to that new site. Ideally, systems can be brought back online in a relatively short amount of time, probably within a day or so.

Let's say there is a primary data centre somewhere and the CSM wants to make sure that by having a cold site, they can recover in a few days' time. The data, and maybe the hardware, are in the cloud and some other physical location, and there's a plan in place for making the switch. If the network goes down or a data centre is destroyed, the CSM can start firing up those systems in critical order in the new place. This will require some installation work for the systems, restoration of data from backups, and so on. Not the fastest option for recovery.

A more expensive option for redundancy is to create a recovery hot site. In a hot site, everything is duplicated in another location that resembles the primary data centre. Data is copied almost in real time from site to site, with dual systems up and running at all times in case something goes down in the primary site. The big benefit of using a hot site is that the switchover to a backup is almost transparent to users, at least in theory. Because the second site is hot, it's running all the time—perhaps with slightly fewer resources than the primary one but functional nevertheless. Of course, this level of redundancy costs a lot of money.

BACKUP, BACKUP, BACKUP!

- Have a backup plan made—how, when and what data?

- Store the backups in a separate physical location.

- If you encrypt your backups, make sure you can decrypt as well.

- Test the backup restoration frequently—once is not enough!

- Check whether restoring backups can be done fast enough for business purposes.

The absolute minimum is to have backups of all the important data in another location, not all in one place. If companies don't do that, they've lost a lot of insurance. There's nothing the CSM can do if the data is lost entirely. But if they have backups somewhere else, they always have a chance to recover. Remember 9/11? A lot of companies went down with those two towers. The ones who had backups from their data and some capacity to rebuild their systems mostly made it through the disaster. But there were many that didn't.

THREATS TO CONTINUITY

Business continuity risks are perhaps the most important ones to prepare for. Losing the company's data is one, of course. It's amazing how many companies don't have backups. Many think they do until they test it the first time, usually in the event of an actual disaster. Unfortunately, some of them find out the system was behaving in an unexpected way. Maybe it wasn't really doing a backup, or maybe the backup was encrypted so that no one could get their hands on it, but the only encryption key was in the primary system. Or the decryption and restoration process turns out to take so long that it doesn't matter. We've seen all of these scenarios happen, and whatever the reason, they happen a lot.

One hospital in Singapore demonstrated just how bad these scenarios can get. They had a problem with ransomware, like nearly every company; nobody is immune, it just varies how much damage the attack does to operations. This hospital got hit badly. They had ten locations running—all hospitals and clinics—and someone clicked a link on a phishing email, downloaded ransomware and got the receptionist's computer infected.

After the initial infection, it took around three months until the hackers launched their attack. They had been infecting everything inside the network during that time. When the ransomware hit them, it activated on seven of the ten sites, and all of the servers and workstations. All of the data in the servers were encrypted. The company received a message telling them to contact a certain email address and pay a large sum of money to get the decryption key to get their data back.

Seven locations had no access to their office records or patient records or anything else. This is, of course, a continuity event. They couldn't bill patients, couldn't see customer data or anything. The IT manager went to restore data from the backups. They expected to be up and running again the next day, or in a few days' time at least. But guess what? Their backups had been done with the same servers and technology that they used for storing the data normally. They had set up Windows shared folders and used software to back up data from servers and workstations to this backup share. The ransomware, of course, was able to find all shared folders in the network and encrypt everything. So all of their backups were encrypted, too, by the same attackers. Data wasn't copied to an offsite location away from the company's operational IT systems. Big mistake.

So they had two options. The first one was to pay the

ransom. The second one was to re-enter all the customer data manually from paper printouts. They finally opted to take whatever printouts they had in their archives and re-enter the data manually. Since they didn't pay the ransom, they were never able to access their data, and they ultimately lost it all.

Whether it's a ransomware attack or something else, the most common danger of IT disruptions is system downtime due to system malfunctions, unexpected changes or any other unforeseen reason. Companies don't understand how often this happens and what it means to the business. They think they have backups, so it doesn't matter. When a problem occurs, it's many times more expensive to fix than the cost of preparing for something up front. The lesson is, if companies do nothing else, they should do backups and do them often. End of story.

SCHEDULING BACKUPS

Nobody wants to risk a total loss of data, so one task that always makes the critical list is to put a system in place for regular backups. 'Regular' depends on the type of business and system. This obviously needs to be planned well because the amount of systems, data, backup frequency, restoration tactic and every other detail has to go in there.

There are two things CSMs need to figure out when

they're planning a backup system. The first is what data is important and needs to be in the backups, like customer data or banking records or encryption keys for the backup. The second is how often to make the backup. This might not seem very important, so many people tell us they do it every Friday and assume that's good enough. But if they're collecting data 24/7, 365 days a year, weekly backups aren't often enough.

What they don't think about, in many cases, is that the date and time of a backup defines the point in time to which they can return—the latest data they can get. Think of it as a time machine that can go back to the moment when the last backup was made. All progress that happened after that moment can be lost forever! A problem on Saturday morning is fairly minor if there's a backup from Friday night. But what if the disruption happens on Friday morning? That means they will lose one full week of data because they have to go back to the last week in their backups. In business terms, that means the company has decided it's okay to lose one week's worth of data. In professional terms, we call this their recovery point objective, RPO. Their RPO is then seven days' worth of data that can be lost, and it's okay to the business. This is a fairly high-level decision in any company. A lot can happen in a week. In most companies, that is not acceptable.

The CSM should advocate for frequent backups. If IT

gets to define the RPO, backups will most likely be made every Friday for all systems, or maybe once a month with an incremental backup weekly for whatever data was changed. Quite often, they forget to ask the executives what is actually an acceptable amount of data loss that the business can bear. The wise cybersecurity manager makes sure that this discussion takes place and that everyone understands the implications.

UNDERESTIMATED LIKELIHOOD

If there's a backup system in place, a lot of people in the company are going to ask the CSM why they should do more than that. People will ask the cybersecurity manager why they should care. They've got this backup system, but they don't see why they need a plan for it. The IT security manager or cybersecurity manager has to answer that question.

Why does the company need a backup plan, or why should they test backups? Why spend on redundancy, or something as expensive as a hot site replication? Because the probability of IT failures and cyberattacks is fairly high, much higher than conventional disasters like fires or floods. These things happen almost yearly, sometimes many times every year. Cyberattacks occur routinely; a fire happens in a business building maybe once in fifty years, on average. Yet most of the office buildings are mandatorily equipped with fire exits, fire alarms, sprinkler systems and so on—often much more expensive but invisible to the eye once people overlook their importance.

UNACCEPTABLE RISKS

Remember, there are risks that you or the business can never accept if they are in their right mind. Continuity risks are often like this. Nobody would risk the lives of employees, family or themselves. Why would they jeopardise their whole business if they understand the risk correctly? Of course they wouldn't!

BUSINESS IMPACT ASSESSMENT

One effective way to demonstrate the need for a plan is to do a business impact assessment (BIA). Doing a BIA lets people see what would happen in different scenarios. For instance, the CSM could work with a technical person and a business person, and challenge them to imagine what would happen if a certain IT service or system was not available. What if it's down for a couple of hours? A day? A week? What happens to the business? By considering these scenarios, the participants will see that the longer the downtime, the bigger the impact on the business.

In most cases, in two hours, probably not that much happens. The service desk handles the customers with apologies. What if the outage lasts a week? How would people respond? Is there any way to do things manually for a longer period of time? What plans should be in place for this situation? Writing these procedures down and estimating their cost and effectiveness creates a BIA.

The BIA can be extremely useful, but most companies don't do it. Even companies that recognise the critical nature of certain systems don't usually do a BIA.

ASSESS IMPACTS QUICKLY

- Find a few people who know business and IT.

- Make a one-page form where you can write answers down.

- Ask them what happens to the business if a system is unavailable for a few hours, a few days, a week or a month.

- Record the result of the discussion on your form.

- Done!

- Repeat for all important systems.

Companies usually begin doing BIAs when they grow to have hundreds of IT systems in place, and they need to classify them and decide which ones are really critical. A BIA can help them do that. At first, they may list thirty to fifty items as critical, but after doing BIAs, they might be able to narrow it down to ten critical systems. Not everything can be considered critical. The BIA has the power to help companies prioritise.

DEVELOPING DISASTER RECOVERY PLANS

Sometimes, all the preventive measures fail, and disaster happens. That's when the company needs a disaster recovery plan, or DRP. A DRP explains in detail how to recover from certain types of disasters. For IT, it should be technical in nature, like planning for blackouts or disruption of critical IT systems. A company can have a bunch of these plans—one plan for each scenario.

One scenario could be losing network connectivity—not having access to resources at your main data-processing site or the internet. For example, this could involve losing connectivity via VoIP to your customer service, or something like that.

For the plan to be effective, it should be drafted with the people who are actually responsible for using it when a disaster happens. The cybersecurity manager cannot make this plan all alone. Ivory tower documents don't matter at all if an IT disaster strikes—but this DRP should. The cybersecurity manager should think about who will react and take action to bring the facility or system back online. Those people should be on the team that creates the plan. It's also critical that the people who are actually responsible for bringing systems back online have access to the DRP, have paper copies of it, and know how to use it.

The CSM doesn't have to be too pushy about getting these people involved. They just have to say something like 'Hey, I have these headers here on an empty document, and I need to explain what to do in the event of a disaster. So I'm going to share it with you online in our collaboration platform and ask you to fill it in. I'll send a printed copy; then everybody can access it if something happens. Sound good?'

The DRP doesn't have to be complicated; it might be expressed in simple bullet points. In minimum, it should contain a list of systems in order of importance, along with a list of instructions about what to do and whom to contact in case of emergency. The plan should also have actionable details, like how to run diagnostics to find out the cause of the problems, log in to systems, restart servers or services, rebuild entire systems, restore data and so on. It's a very different type of document from a policy. A policy should be readable and understandable. The DRP should be useful for the people who are restoring the system—step by step, in very succinct terms. Its core purpose is to be there when people are working under a lot of stress and they have no time to start looking for instructions. They're going to be working under a lot of pressure, so if the plan helps them find the information faster, then it will work.

The steps in the plan might be something like 'Is it working? If not, run this command. If that's successful, it's working. Go to the next step. Sign in to this system. See a log and what's in there. Stop and start the service in question.' Very technical, very straightforward. A trained professional might do this in a few minutes' time once they get a hold of the system and the DRP document. In the best-case scenario, at least a few people will have access to the plan and practical experience on how to restore normalcy after a disaster.

With a concise, well-organised DRP, the company might be able to restore operations in a matter of fifteen minutes after a minor disaster. Without a plan, it can be a huge circus just to discover where the problem is and find out whom to call. If too much time is spent trying to sort out who's in charge, people will start to do their own analysis without understanding how to debug the problem, and that makes things even more difficult. In such cases it often happens that people make incorrect deductions about the cause of the problem, and the fix will take even longer.

But if all the commands are listed in the DRP, anybody on the team can log in and follow the plan. There's no need to be dependent on just one person. Anyone with a copy of the DRP can execute it.

TOO MANY BULLDOZERS

Here's a story that illustrates the key points in this chapter. There was a payment processing company that processed credit card transactions for almost an entire country. It was a huge component of the economy in that country. If the payment processing company were to fail, people would stop paying in the stores. Failure was not an option.

The company actually planned pretty well. They had two physical locations for their data centres that were close but not too close. All the systems were redundant—they just duplicated all the servers and all the data; it was thought out and executed well. But they didn't test several scenarios, like what happens if they physically lose connectivity between those two sites. Would the customer still be able to pay, or would the entire payment processing operation go down?

In between those two data centres, a construction site went up. The construction company dug a hole in the ground, twenty metres deep. One of the bulldozers cut the fibre line between the two sites, and the payment processing company's primary site was disconnected from the internet. Then the switchover to the secondary site didn't work right, so they were down for the remainder of the day until they could figure out and bypass the problem.

In another situation, a big internet service provider and mobile service provider had a similar problem. This was also due to a bulldozer, but it was way worse. They actually thought that their two physically separate fibre optic lines in two locations didn't go into the same service tunnel, but they did. When the bulldozer hit, all of their customers were cut off and couldn't be restored until the physical fibre optic line was repaired. This put down communications for hundreds of thousands of people in the metropolitan area, and all of their business customers as well.

Redundancy is often a challenge in communications because companies often buy communication lines from two or three different internet service providers but don't realise those companies are sharing one physical fibre. If that line is severed, having a variety of providers doesn't make any difference. This is especially true for transoceanic fibre lines—there just aren't many choices.

CHAPTER NINETEEN

SECURITY LEADERSHIP

Cybersecurity managers must understand how company leadership is organised around security. They have to understand how the company leadership structure works and recognise that security is everyone's concern—not a separate function relegated to a select few. Of course, in larger companies, it's natural to hire specialists to work in various security functions. But in many companies, the CSM can be a standalone role or even a part-time employee.

SECURITY LEADERSHIP MODELS

Security leadership is, first, about who is making the calls. In principle, that means there has to be a structure in place for decision-making around security issues. In reality, what often happens is that a CSM starts with a company, and everybody rushes in with the problems that nobody else was able to solve. The underlying thought seems to be that if it's nobody else's problem, it's probably security's problem. Remember earlier when we talked about risk registers and how they easily become dumpsters for everyone's problems? It's kind of like that.

There are two ways to handle security leadership more effectively. The first model of security leadership is centralised management. The centralised model is when you have one person or committee with oversight and approval of all security responsibilities in the company. It's a one-stop shop.

The second is the decentralised model. This is where a number of different people in various departments within the company each make their own decisions about security based on the needs of their particular area.

Larger corporations might use a mix of both models. A company with many subsidiaries, for instance, might find it impossible to run security in all of them from a single centralised security team. People working in the subsidiaries probably take instruction from headquarters hundreds of miles away with a grain of salt. They might complain that the home office staff doesn't set foot in their facility or do the hard work, that they just issue the rules. Frankly, quite often, the staff in subsidiaries don't even know who owns their shares, let alone who is their counterpart in corporate security matters! In those cases, it might be better to use a hybrid model—employ an on-site security champion or an agent who learns and understands the local operation but who is also cooperating with the centralised function.

We worked with a company that had about forty thousand employees in around seventy standalone subsidiaries and three independent business divisions. They set up a centralised security function for managing all of the security in the entire organisation. They had around five people in the centralised function and then a large number of roles doing different, separate security tasks around the different locations. There were working groups for different security issues; for example, health and safety alone had more than seventy people working together. Similarly, security in IT employed more than a handful of full-time professionals. Of course, they also held a lot of conference calls and virtual meetings where people could discuss things and follow up with programmes and policies.

In a centralised model, the advantage is that you have all the decision-making power in one place. They can easily get policies drafted and accepted with the headquarter executives. It's fast to make decisions, and you don't suffer from the confusion that can set in when ten different people are doing things their own way in separate parts of the organisation. From the company level, the centralised model looks more organised, and that's a CSM's dream—to have all things well organised.

The disadvantage of the centralised model is that those sitting outside of that central function, far away in a branch or in a subsidiary in a different country, don't have much contact with the central security function. They're left pretty much by themselves with little support. The policies, procedures, and guidelines that come from the ivory tower might be good for their part of business or totally unsuitable, but once polices are issued exceptions, it will cause trouble. This can create friction. Plus, dealing with security is time-consuming. The time the home office spends doing centralised security functions takes away time and focus from what they should be doing—running and growing the business.

The advantage of a decentralised model is that the authority and discretionary decision-making power given to different branches of the organisation allows them to create the best processes for their unique situation and circumstances. There is no omnipotent ivory tower issuing policies that don't make sense for their branch. Perhaps, the headquarters just issues a high-level policy that the rest of the cybersecurity has to be managed so and so. Call it a meta-level policy that sets a requirement for subsidiaries to build their own security management systems. The disadvantage is, of course, that policies can end up being wildly inconsistent, and security standards may vary from one location to the next.

PLAY ALL THE ANGLES

A savvy CSM will have to approach cybersecurity management from at least four different angles:

1. Policy

2. People

3. Systems

4. Third parties

One detail we like in the decentralised model is that it allows things to get done, and it is very adaptable. The branches know what they need, so they do it. In contrast, people in the ivory tower seldom get their hands dirty. When they do try to go hands-on, it's often a disaster. Because they don't know exactly what the branches do, the people from HQ may end up creating misguided or inefficient policies.

Whichever option you decide to go with, centralised or decentralised, try to consider how to take the best of both approaches. The old wisdom in Zen is that best way to keep a herd of sheep in control is to let them loose on the pasture but keep a vigilant eye on them. You will need to give a certain amount of localised freedoms, but you also need to be the person who makes the final call in all things that affect the whole enterprise.

SECURITY LEADERSHIP DIMENSIONS

The CSM should look at security leadership from several angles—leadership is about many things, not just about managing security matters.

This is the high-level leadership framework that we've used successfully in many of our customer engagements. It is easily understood by high-level management and covers most of the cybersecurity leadership issues from a management point of view.

At the policy level, a CSM would make sure that all dimensions of cybersecurity that are relevant are governed through a suitable policy. From a people perspective, he should engage the teams, governance models and executives who are key in making the policies happen. Then those people and their security responsibilities would be included in the relevant policies and approved by the top management. From a systems perspective, the CSM would see that this cybersecurity governance structure is enforced by security technologies and that the key technologies will be included in the policy framework. Finally, he would see that third parties, like vendors and service providers, are covered through requirements in their contracts.

SUPPORT THROUGH POLICIES

Policies are important to the CSM. Without security policies approved by top management, the CSM has no authority over anybody or any issue in the company. Without management authorisation and support, the security manager is just another employee.

The highest-level security policy in the company needs to explicitly state the responsibilities and authorisations of the CSM's role, goals for cybersecurity and management support for it. The policy should define things like its scope, purpose and goals. The scope describes all the things that are governed under the policy. The goal is the purpose and objective of the policy—it has to make sense and has to have a reason to be there. Each policy has to be for somebody and about somebody. The highest-level policy for cybersecurity might just state that cybersecurity is essential for protection, and it supports the business goals of the company.

The 'what' part is the actual requirements of a policy—

usually called policy requirements and stated in plain English. These must be stated clearly in the policy so that anybody can understand them, even someone who doesn't speak security lingo. It's best to make the highest-level policy short and understandable, a statement from the management more than an operational document with a lot of instructions and requirements.

GET THE RIGHT APPROVAL

Once the CSM has formulated a policy, it needs to be accepted by an appropriate level of management in the company. The highest layer is the board of directors, the biggest big shots. They represent the shareholders and sit above the top management of the company. To get a policy accepted by the board, the CSM has to work with the CEO and management team to get the policy on the agenda for the board to review it.

The board's responsibilities include oversight on the largest risks in the company and oversight on compliance issues. For example, in some countries, it's mandatory for financial organisations to have board approval for risk management policies. The good thing about having such a document accepted at the board level is that management will be obligated to follow it. Obviously, not many policies will ever be accepted at the board level, and seeking acceptance from that body should be expected to be slow. Boards meet only so often.

TOP-LEVEL POLICY

The CSM will also work with the layer below the board, operative management. This is the senior management team, the C-level executives, and the CEO. The CSM is not running every policy for every risk by all of these people; most companies have one overall information security policy document. It's just a few pages long, stating what the company wants to achieve with security, what the main goal is, who runs the show with regard to information security and who's responsible in each department—the policy should state a function or role but not the name of the person.

The information security policy document should also include some general guidelines or goals in different areas. For example, it may state that the company wants to support a new product line and that a certain position title within the organisation will be responsible. 'The VP of software design will be responsible for this goal.' The document should then provide an outline of where security belongs and who is responsible for it in different parts of the organisation. Finally, the information security policy document should state that not following the policies may be grounds for disciplinary action. It should be signed by management or the CEO.

Below the top-level security policy document, there are additional, more specific policies, and these are much more detailed. These include things like HR policy, HR security policy and IT security policy. There may be very detailed policies about each function. Usually, the heads of those functions have the authority to approve and sign them. Sometimes a sign-off by another member of the management team may be required.

WHO OWNS IT?

Defining who makes the calls may be the most important part of the policy. For example, if a cyberattack hits the help desk, there might be uncertainty about what role and what authority the person leading that team has. Let's say there's a spam campaign with phishing messages and malware hits, so people are calling the help desk. Who is responsible for escalating things in that process? Is it the security manager or the help desk team leader?

For most of us, it's evidently the help desk team leader who should handle this scenario, and that department head should have a policy-defining chain of command. A lot of people in the organisation, though, might think it's the security manager and would call the CSM directly.

Quite often, the security manager makes the mistake of starting to sort out the issue because someone called in to ask for help. That's not the proper way to do it, usually. The CSM should direct everyone to the help desk and say, 'We will work with the help desk to address this threat. But the help desk leader is in charge.' If it's a security incident in a large company, the company may have a dedicated incident management team who can take over the investigation and remediation of such incidents. But it all begins with the business function that initially noticed the problem. It's their job to know when things need to be escalated further.

To make leadership decision-making responsibilities clear, a small company might have weekly or even monthly meetings for policy issues, then present them to management when necessary, while a global company needs a lot more structure.

Leadership decisions and policy-making tend to be slow-paced. The CSM should be prepared to reserve a meeting, have it in two weeks, help write out a draft of the policy, go and have it accepted, announce it to people, train the staff and then implement the policy. This process can take months.

Occasionally, decisions get made faster, like when a disaster strikes and the CSM needs to fix something immediately. There's no time to involve any managers. This layer also has to be defined within security leadership. If there's a service desk, and the company is hit with a cyberattack that makes people call the desk, they should have the authority to handle the issue immediately without having to escalate it and wait for decisions from above.

There is no time for management meetings for very urgent issues. For example, there should be no waiting when it comes to phishing campaigns. If the service desk, IT operations or other departments under the pressure don't have the authorisation, they should take the risk to just get it handled. Ideally, these types of urgent incidents should be provided for in written policies and incident management procedures.

WHAT ELSE TO INCLUDE

The CSM should involve leadership in policy creation whenever possible. Ideally, the CSM wouldn't actually be writing the policy, just supporting the creation of it. That means when helping a department create a policy—like HR security or IT security. Quite often, however, it ends up that the CSM does actually write some or all of the policy for the department, and then the department head comments on it and revises it as needed.

Responsibility should be part of every policy. If there is

a policy that doesn't state who it's for and who will be acting within its scope, it's just a paper that doesn't affect anybody. Nobody needs to read it, and nobody needs to adhere to it. The responsible parties must be spelled out in the policy.

Most companies take security issues very seriously. There might be a general non-compliance clause that states if employees don't accept and follow the policies, they might be punished somehow. We've seen companies go so far as to take away the computer from an employee for a period of time if they fail to follow a policy that requires employees not clicking phishing links. It sends a clear message: pay attention and follow the policies. The harshest way to get the word out that the company takes security seriously is to fire a person for a security violation, and it usually happens after a huge mistake is made. We don't say that punishment is the appropriate reaction to everything, but the possibility to use it has to be present. Use it sparingly, though—reward is also a powerful tool.

CHAPTER TWENTY

HR AND PRIVACY

When hiring a new employee, most companies fail to do the most basic security checks. As we discussed in chapter 8, they usually don't even check the applicant's official ID. So really, the person showing up for work could be anybody.

Most companies also fail to verify the applicant's claims. Many, if not most, of the CVs that companies receive don't accurately reflect the truth. They aren't necessarily fake, and there's some truth to most of them, but CVs sometimes exaggerate or misrepresent a candidate's experiences.

For example, we've hired people who told us they were experts in information systems and certified in operating systems and certain technologies. In reality, they didn't have those skills. This happens frequently, and it's not just an HR issue; it might be a security issue, and most definitely, it's an issue about money. If an applicant is fabricating work experience, what else are they lying about? How much money did the company lose just in hiring and firing people? That's a huge financial cost, too, but not considered a security issue often. But it's the same company's money that pays that bill too!

If HR managers don't perform their due diligence at the start of an employment contract, they might end up with questionable people working at the company. So it is essential for HR to conduct background checks consistently on every new hire. Many companies just fail at this outright. They don't even do background checks for security positions.

Failing to do background checks doesn't make much sense. It's relatively inexpensive to run a check, and yet the cost of hiring the wrong person or a person with a criminal background is quite high, sometimes catastrophic. It's so much easier to hire the right person from the outset. HR is critical for getting the right people and avoiding the wrong people.

> Find a professional service provider who can do background checks for your HR.
>
> Security screening is something you can't do in-house.

HR also fires people, and if they don't do it the right way, that can be a huge risk to the company too. Many employees who leave a company take information with them illegally when they go. That might come as a surprise to some people, but it happens a lot, especially in sales organisations. A salesperson's value depends largely on who they know, so the next employer might value them more if they have an address book of contacts. Realising it might give them an advantage, some salespeople try to take the sales database with them when they leave a company—secretly, of course. There's a fine line here. If you have a business relationship with a customer, no one can expect you to not contact them again under a different agenda, but taking the whole sales database without permission is illegal. This risk can sometimes be covered with anti-competition clauses in employment contracts. But in reality, this happens a lot, even with those clauses in place.

We had a CEO contact us and ask, 'Can we track down whether someone has opened our sales database files and made a copy of them?' This happened after they fired a salesperson, then suddenly their customers started getting calls from that salesperson's new employer. It was well timed and looked very suspicious, but in the end, there was no proof. The CEO wanted evidence that there was theft of data. The problem is that most companies forget to enable the file audit logging feature that allows the company to make a log of who opened or copied a file. No log, no evidence. They can't generate logs retroactively.

It's not just salespeople either. Maybe a system administrator left the company with all the data about the employer, like passwords and keys to information systems. Our advice is to be very nice to system administrators when you fire them. They can wield enormous power in terms of access to information and systems. Treat them well, even during the hard times.

PRIVACY GUIDELINES

HR plays another role in security efforts when they're involved in complying with data privacy legislation. This area is quite new. The titles of privacy officer and data privacy officer have just started to appear in the last few years, as new legislation has started to mandate customer and employee privacy rules. HR used to be the lead in this area, but now there are managers with titles like chief data protection officer. Whatever the titles, HR is still at the heart of this because HR deals with employees' private personal data. Sometimes they are needed as in-house experts when dealing with customers' personal data.

The HR director is also connected to a lot of different

security policies and will know company requirements about privacy and the local privacy legislation. They're also the key players in dealing with representatives from unions. If a CSM wants to make an acceptable use policy, they'll have to get pretty deep into people's daily work, and they'll need to talk with union representatives, so they will need to work with HR for both. Let's say the security department wants to install a CCTV system to monitor areas where people work, or maybe they want to track their computer use in some security-related cases. These are usually things that people feel are invasive, or they feel it's against their privacy. The best way to seek acceptance is to go through HR, then talk to union representatives with them and plan the necessary improvements together. Early involvement is the key.

Because HR managers have a high level of security awareness, most CSMs are comfortable working with them. HR knows how to protect personally identifiable information, or PII. When they leave their offices, HR people usually take a lot of care to lock their doors, put all of the papers in drawers and lock them, lock their screens and computers and take other physical security measures. Many of these security and privacy requirements are mandated by law. If HR makes a mistake, they might be liable for it, and they know it.

THREATS IN HR AND PRIVACY

Data breaches and leaked personal information are the clear threats related to HR and privacy. Typical cases include stolen personal IDs or social security numbers. Stolen private data might also include street address, name, phone number and credit card number, which is enough to conduct many different types of fraud on someone. Nowadays, it's becoming very common that health institutions are breached, and all of the above data is stolen along with patient health records.

The simplest form of identity theft is someone stealing personal information and using it for phishing. There's much more, though. Some people's identities have been used to buy or sell a home or to shop online. It's easy enough—some online shops allow customers to make purchases with post-payments, so all a hacker needs is a full set of information about a person's identity, and they can make a purchase and send it to a false address, leaving the victim to pay the bill.

Identity theft, as horrible as it is, is seldom personal. Hackers rarely try to target somebody personally as a vendetta. It's all for money, and there are a million targets out there. The only limiting factor is that attackers don't have the resources to attack everybody at the same time. The one who is targeted is just unlucky. And these unlucky guys come in large quantities. PII is one of the hard currencies that cyber criminals use in their trade.

Once personal information gets out, there's not much a victim can do. We recently heard about a case where someone in Finland was involved in a data breach. Back in 2012, his information was breached along with that of nine thousand others. You might think that information from 2012 was too old to be useful anymore, so what's the big deal if the information was leaked? That may be true for a phone number or an address, but some private information doesn't change, like a person's social security number or their name. In 2018 this victim found out someone was fraudulently buying things online under his name, six years later! Once it's out, it's out.

Theft of personal information obviously leads to a loss of consumer trust, and it should be a major concern for companies that have a lot of B2C consumers. It's getting worse; one recent data breach contained 1.4 billion user passwords. This trend has led countries across the globe to enact new laws about data breaches and spreading information with malicious intent. It's going to take time for these new laws to become effective; however, the criminals won't wait. They'll do it anyway, with or without the law.

Companies can spend a fortune to make IT and technology virtually impenetrable, but the bottom line is that the people involved need to develop their security awareness. We have data that shows the biggest thing companies can do to improve security is increase awareness. Otherwise, someone is going to click the wrong link or open an attachment or do something silly. And there's no patch for human stupidity.

Companies usually perceive themselves as being better prepared against these sorts of threats than they actually are. Employees often think, 'Because I have a computer in front of me and I can do anything I want with that computer, then I can always avoid clicking a suspicious link or opening an unknown attachment. Because I can do it, I deduce that everybody else in the company will do the same.' It's not true. Most people don't understand when they are being influenced or tricked. Human nature is one of the most difficult problems in security, and it always will be.

SOLUTIONS

To protect private data, CSMs should find out where the company's customer data and employee data are stored. These kinds of data have a lot of similarities; they're usually linked to individuals, such as employee name and details, customer name and details, or partner name and details. The business might not even be aware that they have this data. They might focus on the web shop, sales or operational processes, but they often forget the data has a lot of value both for businesses and for criminals.

Often, companies think they just need to secure the payment process, credit card information and application. Sorry, that's not all of it. The customer data within the application and the servers where data is stored are also important to secure. Sometimes that data goes silently into log files without anyone realising it. When logs are stolen, all the data go out too, even if it's not in the customer database anymore.

The CSM should cooperate with HR to find out where the data is stored, then figure out if it's stored securely enough. Are there risks that still need to be addressed? This usually leads to a list of servers, services, IT services and a description of what sort of data is in each of those systems. Then the CSM must meet with HR and IT and have a discussion about how to handle securing that data without stressing the budget.

SECURING PERSONAL DATA

Companies usually decide that they've got two important types of PII data to protect: customer data and employee data. Employee data is straightforward—HR will be responsible for securing records in the salary and payroll systems. HR may need to do encryption or test the HR system's security. It shouldn't be a problem for the CSM to insist on this; most HR staff will be happy to do it because they know how important it is to protect that data.

Customer data is more difficult to secure, not so much from a technical angle but because the organisational roles aren't clear about who owns it. It's not really IT's data. IT is just running the systems. They can secure it, but they'll look for an additional budget to do so. This is when the CSM goes to senior management—maybe the leader of that business operation unit. The CSM will say, 'We've got this service that contains a hundred million user records of our customers, and I'd like it security tested and audited.' The manager will ask why, and the CSM can explain its importance.

Internet-facing systems with customer information face the highest risk. The best solution available is not to store that data at all, if possible. The next best thing is to design the system to be secure from the start, then encrypt customer information and do security testing when the system is ready enough to go to piloting and again before it's put online. Preferably, the testing and consultation will happen early on so developers have time to fix and implement security fixes if necessary. We've done hundreds and hundreds of penetration tests against different web applications. The best results in security tests are consistently with systems that were designed to be secure from the start.

> It's very hard to add security on top of an existing system afterwards. Get involved with the software development teams in the company!

EDUCATING EMPLOYEES

As we've seen, the weak link in securing data is often the employees who lack training in security issues. Fortunately, HR usually has funding budgeted for training and employee education. They organise e-learning and face-to-face teaching and online training events as well. It shouldn't be too difficult for the CSM to get approval for an hour of cybersecurity training for all employees. If possible, scheduling a one- or two-hour face-to-face training for all new employees is ideal.

E-learning saves a lot of time for everybody, but face-to-face training classes are important as well. It's helpful to make the CSM's face known to the other employees. They need to understand that the CSM is a people person who likes to help, not just stay in her office and play with computers.

Those are all good ways to train the entire staff in a big company, but there's usually a smaller need for specialist trainings. This includes people in ICT special roles, such as network specialists, firewall administrators, application developers and many others.

EFFICIENCY THROUGH E-LEARNING

Most companies do this type of training via e-learning instead of face-to-face, in-person training because it's more cost-effective, has flexible timing, and still meets compliance criteria. These trainings live on an internal cloud service where HR posts the e-learning materials. The content is usually created by a service provider who's able to produce the content, like educational videos. It's also possible to buy this training as a service and just modify small bits of the presentation.

Let's say the company is building their own web services platform for their customers. This is supposed to make money in the future. Shouldn't the people who are coding the software and putting it online also know about how to create secure web services? Were they ever trained to do that at school? Probably not. These people are usually somewhere in their forties. When they were young and in school, there was no subject called 'security' in ICT. It's being taught in schools and universities now, but not often enough. When they do address security, universities aren't teaching the most up-to-date tools in a professional context; they tend to rely mostly on textbook theory.

That means the CSM has to be aware of training opportunities and courses for IT specialists in a variety of roles. Then they need to cooperate with those departments and with HR to make them aware of the training or event. HR and the business unit will then discuss who will pay the bill. It's not a hard sell; they'll usually buy a lot of these types of training.

Continuing education is always a good business ethic. And it's an effective way to show people that their biased sense of security is not correct. Unless employees feel that there's urgency and a need to change, there won't be any change in the company.

SECURITY AWARENESS TOOLS

We've talked about the most common awareness training tools: e-learning and face-to-face training. There's another way CSMs can teach and test awareness: by using *phishing campaigns*, which test people's awareness of phishing and related threats.

In a phishing campaign, the company sends phishing test

emails to see whether anyone clicks. Then they make a list of the recipients and note how each responded or didn't. The results give the CSM valuable feedback about employee awareness outside the classroom.

Companies can also test whether employees are vulnerable to being fooled with a service called *vishing*. This test is focused on telephone entry points—usually a help desk or service desk in ICT. The security tester will call the help desk and portray themselves as an employee or manager, then use social engineering tricks to influence that person to give out a password or change it or even create a new account.

This is a useful test because this is exactly how many companies get hacked, unless they are wary of the threat. Kevin Mitnick, one of the first hackers to make the news and now an author, used these techniques. He exploited human behaviour and trust to hack a lot of companies. He was good as a technical hacker as well, but his main tool was always social engineering. If he called in to a service desk and someone answered, 'Hello, Company XYZ, how can I help you?' he might start with, 'You can help me by giving me the CEO's email address.'

SOCIAL ENGINEERING

Sometimes asking directly does work, but usually hackers

are a bit more clever. They use social engineering techniques to fool people. Social engineering is an art and science of influencing people to do things they normally would not do.

LEARN HOW PEOPLE ARE INFLUENCED

Read Influence: The Psychology of Persuasion by Robert B. Cialdini.

Hackers use something called weapons of influence to make people do their bidding. Human nature is rigged to be that way—there are cues that we follow in daily life automatically without questioning it much. Let's review a few examples of how hackers exploit these tendencies. An attacker could call the help desk and use *consistency*: 'Hey, Sheila, thanks for that sales director's email address—you are super helpful! Can you also help me with another issue—I also need to reset my password. If you could just kindly send the new one to my personal email? I am on the road.' In this case, the attacker was using a technique called psychological labelling. If Sheila accepts the proposition of being super helpful, she should feel internally consistent helping the customer again with the password.

Another type of common tactic is *sense of urgency*: 'Can you make an exception here? I'm really in a hurry.' There will be a sense of urgency. 'I'm about to get onto an airplane, and I cannot get into my computer, so I'm sending this from my private Gmail account.' They'll drop in crumbs of information that seem to be the right way to verify things, then people respond to the urgency and want to help.

Reciprocity works as well. If an attacker makes the appearance that he's been very helpful for the victim even in the tiniest way, he can ask the victim to return the favour. Maybe reset the password or something similar.

Another social engineering approach is to use authority: 'Hey, I'm the CFO and am getting onto a plane. It's urgent, can you just help me reset my email password?' Or maybe there's a phishing email from the police department or the security officers at the bank.

Sometimes social engineering comes with a sugar coating. It works well, especially if the victim likes the attacker or if the hacker is using a sexy picture to present himself. Old-school seduction works.

The permutations of possible social engineering hacks are endless, so awareness testing and training requires continuous training and an ongoing budget. People don't change behaviour from one session of e-learning; they only learn with repeated reinforcement and repetition of the experience. Employees need to get bits of security information now, and then that needs to be repeated many times over. It's much better to help them learn this way rather than by getting badly burned by their mistakes.

Even with extensive training, people are still hardwired to fall for social engineering attack techniques. There is no final remedy to this human weakness. Take this into consideration when you plan your defences. For instance, consider all workstation networks to be contaminated and compromised. That approach should give you the right attitude and allow one more level of defence in depth when you defend your networks.

MONITORING SERVICES

Employee-related cyber risks aren't limited to the company offices or systems. Third-party breaches are all too common—employees use their work-related computers and accounts all across the internet. When these outside services experience a breach, employee account information might be at risk—especially if they reuse the same passwords that they use at the office.

There are services that companies can use to monitor if their employee or customer information has been stolen or leaked. These services monitor for any leaked personal information and alert the company if there are indications of a personal data breach. Most of these stolen personal records are from third-party services on the internet, and they are mostly unrelated to the company's own business. How does this affect their security, and why does it make sense to monitor it?

Let's say there is a breach in a third-party service, for example, at a social media platform called FaceGram. The company would be totally oblivious that some of their users were victims in this breach. But if they used the monitoring service, they could know immediately that there's been a recent breach at their online application. In that breach, there could be 190 million user accounts that were stolen, including email addresses and clear text passwords. They could see, for example, that Shawn Lewis Legrand at Adidas is using the password CATHERINE. This one company could have up to thousands of affected users without knowing if they don't use this kind of a service.

Shawn Lewis Legrand might be using the password CATHERINE at the company IT systems as well as for her work password. Or if it's a social media password, someone could log in to the profile and use it as a means to abuse trust, like the phone call to the service desk—using the personal profile to contact someone and influence them, or ask for a password reset. Suddenly, these two unrelated things become a serious threat to the company.

Remember how easy it is to use Google to find all the login forms for a company's IT systems?

Companies should adopt this kind of a monitoring service to see whether any information regarding people they employ or their customers has leaked out. These services can monitor any PII-related terms, like email addresses, credit card numbers, social security numbers and so on. If the company knows what's going on, they can warn a person and educate them about the breach. They can ask that person to pay attention, change their passwords and watch out for phishing emails in the future. They can also educate users not to use their work email in outside services that aren't related to work. After all, the domain name in the email address points directly back home. It's like writing your home address on your key fob. If you lose that key fob, you can rest assured everyone knows where to try the keys on!

Just like social security numbers, once this information is out, it's out for good. If companies don't subscribe to a service to receive the feed of those compromised accounts and records, how do they know whose password to reset? How do they know to warn people that their identity is at risk? They don't. That puts them in a position of elevated risk.

> Monitoring services allow companies to react quickly to possible leaks of personal information and minimise possible damages.

NO INSURANCE FOR STUPIDITY

Here's a telling example of what we're talking about in this chapter. A major company that provides mining machinery and equipment across the globe works in different time zones with a vast network of suppliers who have strange office hours. They have emails and faxes and messages coming at all times of the day. They're large enough that they don't even know all of the companies that provide services for them in their network, though they do know the biggest ones. The machinery might be sent overseas to a site for a customer, then the subcontractor or service provider in that country provides operational support. It's customary for money to flow through the customer, leaving a large bill on the order of several million dollars.

All of these transactions were cleared by email and phone calls, and many legitimate calls and emails happened at weird times of the day. It wasn't uncommon for bills of $17 million or $20 million for big machinery to get approved this way, so employees were used to it. One evening, they got an email with an invoice from a known supplier that looked legitimate and was sent to the CFO, asking for over $18 million. The bill was referring to a known project, and the sum of money was what they were expecting on the bill. The email was forged, but the hackers had done their research. They knew the customer, the service provider in that country and the person in the company who would handle the bill. They had even done their research on the amount to be billed and got that detail right too. They simply added all of those details into a bill template that looked just right and sent it.

The invoice was not questioned because it was expected. All the hacker had changed was the account number. The payment was sent to the hacker's bank account.

Over $18 million!

The money was never found. They had to make a public announcement about the loss because they were a listed company, and it affected their financial outcomes and numbers. It had to go in the annual report, and it circulated in the news. The incident was a massive embarrassment.

There's no insurance for stupidity. You can't insure against careless employees getting tricked by ruthless hackers. You can't insure against an employee giving out millions in free money to crooks. But you can increase awareness and minimise the risks.

CHAPTER TWENTY-ONE

INFORMATION AND ASSET MANAGEMENT

Security tools and increased awareness are key to keeping a company secure, but many companies also need help with a basic question: what, exactly, are they protecting? Many companies simply don't know what they have and what they need to protect. When we do asset discovery services for companies, we almost always find a lot of assets that belong to the company without the company being aware of them. In one instance, the company thought they had way fewer online systems and related IP addresses than they actually did. Upon discovery, we found one-third more IP addresses and related systems than what they thought should be there. That's forgotten, new, undocumented or lost live IT systems that belonged to the company—hundreds of live systems—and they didn't have a clue. Most importantly, these systems were not recognised as their assets and were not managed.

> If you can't manage it, you can't secure it!
> Conversely, if you manage it well, making it secure is easy.

In this domain, companies deal with assets like computers and information. These assets are critical to the business because they create revenue or shareholder value. If information or assets are stolen or damaged, the repercussions can be devastating. Organisations can fail if they don't do this well. Not only in the security space but also in terms of business success. Organisations can incur major losses if they don't get this right.

Because these assets are so important, companies must keep track of where these things are and how they're being used. Information and asset management is not just a matter of having a list of everything the company owns—it's also about how those assets are managed.

For example, we learned about a medium-sized city in Europe that leased their computers instead of owning them—around twenty-two thousand machines, and they were paying hundreds of dollars every month for *each* of those computers in leases. That's not a small amount of money. They used an infrastructure vendor who had a process for delivering new laptops and dismantling the old ones, and they handled the whole process for this municipality. When someone started working for the municipality, he received a new computer from the vendor. When they resigned, the laptop was either scrapped or traded to another user. In theory, this is an asset-light way to manage the end user devices in IT.

The problem here is that the vendor did it all. The municipality oversaw only part of the process, which was fine; it made sense to externalise most of the service. But they also externalised control of the service. The vendor was controlling the number of laptops and the inventories. This went on for years.

What could go wrong? Plenty.

When the municipality did an audit to look at their inventory of computers, they were surprised to learn they had no inventory of the laptops at all. They didn't know how many computers they had or even if they were being used or not. They had no way to verify whether each of the laptops was still there, and in many cases, no way of knowing which user was using which device.

When they finally did the inventory audit, the result was horrifying for the city. Out of more than twenty thousand laptops and computers that the city was paying for, about ten thousand did not even exist anymore. They were just gone. Nobody had any idea where they went. Think about that number—ten thousand computers disappeared. That's ten thousand laptops with hard drives containing potentially sensitive or private information that had gone missing.

The saddest part, with the biggest impact, is that the city had been paying for those ten thousand nonexistent computers every month. That's a huge amount of money wasted. Taxpayers were not happy. Of course, there was a lawsuit, a public case that inflicted a massive hit to the city's reputation.

Asset management is a big deal. It must be done properly. Even if an organisation outsources their asset procurement and leasing to an outside vendor, management and inventory of those assets must remain an inside job.

KNOW WHAT YOU HAVE

Simply put, if you don't know what you have, you can't protect it. Information asset management means being aware of your assets and understanding how to protect them. There should be some kind of policy—you can call it information management, information security or asset management; the name is not important. The policy should state what the company wants to do with all its assets—how to acquire them, manage them and dispose of them. It should include a process for verifying, auditing, tracking and acquiring assets.

> If you don't know what you have, you can't protect it.

The policy should cover laptops, servers, information, acquisition of services, software and licensing. Many of these are physical assets that you can see, feel, use and touch. Companies usually can handle inventory of physical assets but often have a much harder time dealing with virtual assets, like server space, software, licences, customer records, and databases. These are less tangible assets, but they must also be covered under asset and information management.

PROTECTING PHYSICAL ASSETS

Let's start with physical assets—things like servers, laptops, tablets and mobile phones. Many companies don't know what physical assets they have. It gets more and more complicated to track these things as more people use their networks and personal devices, like iPhones and computers, for work-related communications. In many cases, the company might not even own the devices that their systems are running on, so they don't have the right to mandate what should or shouldn't be done to secure them.

PERSONAL DEVICES

Companies actually prefer not to own these devices. If a physical device is required, they'd rather let the employee buy it themselves or lease it for the employee. Leasing means you have to manage the service but not the asset, so it's an attractive option, especially in smaller companies. Leasing, however, doesn't release the company from the responsibility of understanding and managing what they have.

One solution is to use a mobile device management service. In these plans, people who bring their own cell phone to work and want to use it for work tasks can join a mobile device management service that creates an asset directory. That gives the employer certain device management rights, such as the ability to wipe the devices if lost or stolen, and to configure security settings. If you plan to go for this option, make sure that your users understand the fine line between their personal and work use, and that they are giving away some of their rights when they use personal devices at work. Get this in black and white. It's probably a good idea to talk to the legal advisor about how to formulate this on paper correctly. Invasion of users' rights may backfire if the company goes carelessly into this space.

IT ASSETS

IT has traditionally been effective at managing their IT devices. They're used to having racks of servers and piles of laptops, and they usually know who uses what equipment. They just need to make sure to put that information into Excel sheets.

In the past, there used to be just a few servers in a whole company, so creating a list of servers with information about what was in the server and where it was located was not complicated. Nothing more was needed. Now there's a multitude of different devices: servers, workstations, mobile devices, private employee devices, and so on, and there's no single role in the company that can manage them all and the information contained therein. These devices contain all company information, along with any personal information on customers and employees.

Perhaps there is one quality that we'd like to highlight in this context—it is punctuality and capability to organise. People who take care of IT assets in companies should have both. For example, we've seen many system and network administrators who oversee a large number of devices across the enterprise. Not all of them are both punctual and organised, and these personal traits are reflected in what their infrastructures look like. These kinds of people like to keep things tidy, tolerate no mess and naturally like to make lists of things like servers and

IP addresses and keep everything in check. The flipside is that other people might find this attitude as nitty-witty and not very helpful. But believe us when we tell you that this is for the best of the company's cybersecurity.

> If you want your IT assets managed well, hire someone who's both punctual and organised.

LABELS AND PRIORITIES

Knowing what the company owns—servers, computers, information—is the first step. Classifying and prioritising those assets is the next logical step. Putting security labels on information and documents is an old practice that has its roots in defence and government practices. The idea is that once a label is put on a document, the reader should know how to handle it.

Most commonly, companies opt to go for three- or four-step labelling categories. We like simplicity because it's easy for the users to remember what the labels mean. An example of such a categorisation system would be:

- Public
- Internal
- Confidential
- Secret

Want to make it simpler? Just drop out the last 'Secret' category. People often find three categories easy to grasp.

Those labels may be helpful in communicating to the reader how important the information is in each document, but the terms aren't clear to everyone. Employees often have difficulty understanding the difference between internal documents and confidential documents. They aren't sure which documents should be sent to employees and which to partners. They can't decide whether to give an internal document to a consultant. Even top executives and security managers struggle with making a distinction between what should be labelled 'Confidential' and what's 'Secret'.

The best solution is to have a simple rule about labelling things like documents, emails, faxes and other information. Give employees clear if-then statements: If you're giving internal or confidential documents to somebody, make sure they have a working non-disclosure agreement in place with the company. If you create documents and give them to someone else, label them yourself. If the document has confidential information, place the word *Confidential* in the header or footer. You've probably seen the disclaimer clauses in many official email footers. It's the same thing but with a bit of legal jargon to spice it up.

A simple trick is to make official document templates

in Word and PowerPoint with 'Confidential' or 'Internal' labels already present on the documents by default. Then, when employees use the company's official templates, they'll open up with the default label already there. People will make mistakes anyway and won't pay a lot of attention, but at least you will have a default label on the documents. One more idea—add a unique, innocent-looking identifier to the footer of your document templates. Something that's unique to only you. If there's a data breach, you'll be able to search for that identifier by using the breach monitoring services we talked about earlier.

The manual process of labelling information will fail at some point. Any information that's crucial for making money and growing the business, or trade secret information that could be harmful if leaked or damaged, should be secured pretty well. That costs money and time, of course, and should include things like data encryption if it's being sent over the network to other people. You might want to look into things like email encryption, data loss prevention and related technologies. Yes, more securely stored data means that it's going to be harder to access, even for the people who actually need it. That's the double-edged sword of securing data by encrypting it.

LICENCE MANAGEMENT

One more thing that companies should keep track of under information and asset management is licence management. Many products and services that companies use are governed by a licence agreement. If that licence agreement is violated, there could be serious consequences, including fines and penalties.

Unless a company practises good licence management, they may be overpaying for too many licences or underpaying for too few. If the company is underpaying, a whistle-blower could turn them in and get a reward. Say you're working for a major corporation that you know isn't paying their Adobe Illustrator licences for five hundred users. Whoever tips off Adobe could get a fat cheque in the mail as a whistle-blower. And yes, there are service providers who sell this as a service and pay for the whistle-blowers.

Companies should monitor their licences diligently. Treat them like other assets in the information and asset management system.

A MARKET FOR SECRETS

Good asset management extends to the point when a company gets rid of old assets. It's not just buying and managing what you own but also about destroying it securely after you don't need it anymore.

There is a market for secrets, and criminals use all kinds of illegal means to access and obtain secret information that can be sold. In the news recently, an e-waste recycling company got busted when it turned out that certain employees at the company were stealing hard drives out of computers that were supposed to be destroyed, and then harvesting the data for sale.

There are also many stories of dumpster-diving criminals who steal paper documents out of the trash. This happens to many different types of organisations—like healthcare companies who were dumping customer records with medical info by the lorryload, until someone found them. Dumpster diving is still very effective in this modern age.

Security against these kinds of attacks doesn't have to be complicated: get a decent shredder for a thousand dollars. Pay for a good one. Not only should it process without jams, but it should process CDs and DVDs, plastic binders and paper and do cross-shredding. It should be up to the task with a big container suitable for commercial office use. Place shredders where paper is used a lot and where sensitive data is being handled—where people use the papers—printer rooms, the offices of the HR department and so on.

To guard against hard drives getting stolen from the e-waste recycling plant, use full disk encryption on all devices—mobile phones, laptops and desktops. After the computer boots up, the first thing it does is decrypt the hard drive. If the thief doesn't have the password, he cannot read the data. If he removes the hard drive and tries to read it with a special device or another computer, all he'll see is scrambled data.

Encryption of the entire hard drive can be enabled in Windows, on Macs or on Linux operating systems quite easily. Hard disk encryption used to require specialised software, but it's now available out of the box and simple to put into use. There is no reason not to use data encryption on all of the devices your company has. Have this requirement in your policies and include steps to do it in device management procedures.

Finally, we have an example of a company that was leasing printers. Modern printers have mini servers with operating systems and memory; quite often, it's a simple Linux server. Every time someone prints a document, that information is stored in the printer, including things like employee agreements, business agreements, lists, graphs, trade secrets and internal presentations—just to name a few. And that information will be stored on the hard disk of those printers. This company was leasing printers, and once the lease ended or they replaced a device, criminals stole the information from the hard drives of those print-ers. None of those devices had any encryption or data wiping features on them. The data was still there. (Data that is deleted from the hard disk is not usually gone. It's just marked as free space. Getting the data back is just a matter of reading the disk with specialised software and restoring the files.)

An organised criminal gang could go into the recycling business, start buying used printers, then harvest valu-able information off the hard disks. The thief could be anybody; it might be a custodian or a maintenance person, or anyone who has a hobby of hacking informa-tion and frequenting dark web forums looking to trade secrets for money. And this has happened many times over already.

There's truly an economy out there for secrets.

CHAPTER TWENTY-TWO

IT INFRASTRUCTURE

Companies must also protect IT infrastructure, including cloud services, internally maintained servers and related equipment, computers and other internal and external networking resources. From a security point of view, IT infrastructure used to be like a medieval castle with a wall around it and one big gate for entry and exit. The gate controlled traffic in and out, so the soldiers had a single choke point to monitor and control access to the castle. It was pretty safe from intruders.

Today, in a multivendor environment where workers access data remotely and through the cloud, that solid wall around the castle now has a hundred doors in it that all grant twenty-four-hour access. With so many access points, it becomes impossible to make the fortress impenetrable.

Customers might have many ICT vendors, who usually maintain their services, as well as cloud services. Everything's in one big mesh network like a spider web where all nodes can be connected to others. Employees are also using third-party services semi-professionally or just personally, and those aren't controlled by the company in any way. As a result, the old fortress model has been breaking down for years. This is the IT infrastructure world we're now living in. It's a whole kingdom to control and secure, but there's no solid perimeter wall.

THINK LIKE THE ATTACKER

We can barely scrape the surface of how attackers think in this book since our topic is mostly about how to be a successful CSM, and we talk a bit about how to work as a defender. But understanding the enemy is paramount to success. The CSM needs to hold in mind a few key facts about attackers and their mindset. They tend to think and feel quite differently from how IT people feel. Let's compare.

IT cares about people and the company. They want to make information available and its usage easy—attackers want to steal it, destroy it, or make money off of it. IT wants to deploy new services and tools quickly and easily, build new things, and in the process, they make the company more visible to the internet and attackers.

Attackers take advantage of anything visible from the company and use it against them. They have no intent to make things work, only to make the technology and people work the way it was never intended to work, for their own benefit. IT needs to be successful in defending every day, in every single service and with every single person using the services. Attackers only need to succeed once to penetrate the defence. Evidently, the mindsets are opposite to each other.

MIND THE COLOUR OF THE HAT

White hat: Someone working with permission from their customers, using hacker techniques to test applications, networks and systems. These people are often penetration testers working for the company or its security service provider. White hats act ethically when performing tests and reveal their findings only to the customer.

Black hat: People who don't care; they skip ethical considerations of hacking. They might hack or test your systems without permission and reveal your vulnerabilities to the public for personal gain, fame, money or other reasons.

Grey hat: Somewhere in between. Usually, grey hats are people doing things that aren't necessarily ethical. They might do things that are considered controversial, even sometimes evil, but they tend to remain within the limit of the law. Quite often, there aren't laws governing them, so they do it because they can. Grey hats definitely violate ethical standards at times and cause some heated debates by doing so.

The CSM has to understand how to think like the adversary and wage his defence battle against the attacker's mind, not just the technology that the attacker is using. When doing so, the CSM will need support from professional penetration testers, also known as white-hat hackers who help to bring in the attacker's perspective to the game and help the company to test their defences. One bit of advice, though—pay attention to the professional ethics of who you employ to do your security testing. Crooks can't be trusted.

When planning your penetration tests, consider the coverage of the testing you've done. Have you covered single points in your infrastructure, like web applications, or did you cover it all, performing a full-fledged red-team exercise that inspects your whole infrastructure and tests all your defences in depth like real attackers would? Perhaps there are gaps in your understanding that need to be covered by further testing.

THE KILL CHAIN MODEL

The term *kill chain* originates from the military and relates to the structure of a cyberattack. Defenders can use this model to break the attacker's chain of actions during the cyberattack. The model isn't perfect, but it can be useful when the CSM tries to assess whether his company has overlooked different stages of attacks and related countermeasures. There are many variations of the kill chain model. Our simplified model is presented below.

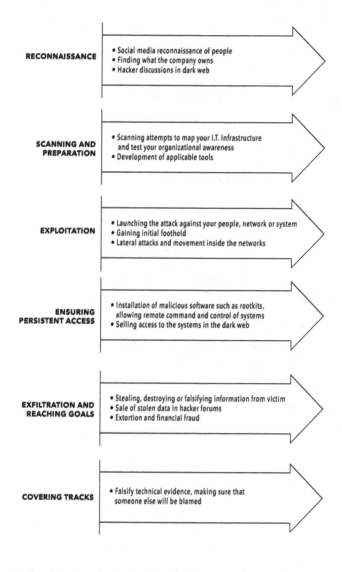

RECONNAISSANCE
- Social media reconnaissance of people
- Finding what the company owns
- Hacker discussions in dark web

SCANNING AND PREPARATION
- Scanning attempts to map your I.T. Infrastructure and test your organizational awareness
- Development of applicable tools

EXPLOITATION
- Launching the attack against your people, network or system
- Gaining initial foothold
- Lateral attacks and movement inside the networks

ENSURING PERSISTENT ACCESS
- Installation of malicious software such as rootkits, allowing remote command and control of systems
- Selling access to the systems in the dark web

EXFILTRATION AND REACHING GOALS
- Stealing, destroying or falsifying information from victim
- Sale of stolen data in hacker forums
- Extortion and financial fraud

COVERING TRACKS
- Falsify technical evidence, making sure that someone else will be blamed

Defending the IT infrastructure against a cyberattack is best done early in the chain. Select defences in your IT that identify and block attacks at each stage of the kill chain to improve your chances to be successful in your defence. When you've figured out which technologies, processes and services are a good fit for your defence, that will be the defence plan. We call it cybersecurity architecture.

IT AND SECURITY ARCHITECTURES

When we refer to IT infrastructure in this book, we're talking about creating a set of written diagrams or blueprints of how IT infrastructure is built. It's like an architectural diagram, and quite often, it's a bunch of technical diagrams due to the sheer complexity of systems. In fact, the goal of the IT infrastructure document is to explain in simple and understandable terms how IT serves the business, what services it uses to do that and how they are interconnected to each other. It should show what the company itself is hosting, what's in the cloud and what resides with external third parties.

A useful blueprint will also contain some sort of depiction of data flowing back and forth—at least for the most important parts of IT. It will show where CRM data is stored, how it is accessed and how it is communicated through the network, servers, firewalls and so on. It should depict business process data flow. In a bigger company, this infrastructure document might be quite complex. But without this understanding, it's difficult to explain to anyone how IT works.

Infrastructure can be explained and depicted from different angles as well. One picture or diagram might represent the network layout, a second diagram might focus on the server layout and a third might depict the data flow. There's no one correct way to describe IT infrastructure because it's a multifaceted entity.

IT infrastructure is a huge topic, and protecting it will require more than this book can cover. A cybersecurity architecture blueprint is a similarly written set of diagrams (and related documentation) that explains how the company defends itself against cyberattacks. It's often sort of an overlay blueprint that supplements the general IT architecture of the company. The kill chain model discussed previously can be used to elaborate on different types of threats, and applicable defensive systems can be set up to counter the threats within the IT architecture. For example, if a company is defending against phishing attacks that infect workstations, they might install an antivirus solution on their workstations and email gateway, and perform automatic Bayesian email spam filtering on their email gateway. They might even go as far as using a technology called sandboxing on their workstations to isolate the email application from other parts of the operating system to prevent exploits from working against their computers. There's plenty of choices for defensive technologies that apply for different parts of the kill chain.

The goal of cybersecurity architecture is to explain in understandable terms how security technologies are implemented in the IT architecture of the company and which threats they are countering.

It's paramount to understand your IT architecture in order to be able to create your overlaying security architecture. Study how your company's IT is built, work with penetration testers, perform all-encompassing red team exercises, learn about the latest attacker tactics and techniques and synthesise your own cybersecurity architecture to counter them. A well-documented cybersecurity architecture shows that the CSM understands his defences, knows his enemy and knows how to win his defensive battles.

DEVELOP A GOOD UNDERSTANDING OF YOUR IT ARCHITECTURE

Observe it from the attacker's perspective and use a kill chain model to visualise how an attacker perceives your assets. Find attack techniques and tactics that haven't been covered in your defences yet.

IT INFRASTRUCTURE SECURITY POLICIES

Developing an effective IT security policy must consider many different facets. In our experience, most companies need to have at least a minimal set of policies in place on the following topics:

- Security of communications
- Networks and firewalls
- Workstations and servers
- Mobile devices

These policies might be short, and they would state in high-level management language what's required from a cybersecurity point of view. IT security policy usually must be approved by the IT director or even top management, so get the structure of the necessary policies designed early on to get them approved on time. Any chapter from this part of the book will be a good topic for a policy. Think critically about which policies you actually need to support the security of the business and which ones would be excessive dead paper.

In addition to these policies that set goals and requirements, there might be a number of technical policies about settings and configurations for everything run in IT. These are often called baselines, and they contain good, known configurations for different kinds of systems that the company uses. Basically, this is a matter of standardisation. When you define what baselines should be used across your systems, you're standardising security in all similar systems across the whole company. This is by far the best way to ensure technical security in your infrastructure—it saves time, effort and money, and it's been proven to be very effective.

CENTER FOR INTERNET SECURITY (CIS)

CIS provides baselines for different kinds of technical environments—like the Windows 2012 server, a version of a Cisco firewall, Windows 10 and much more. Companies can use these policies as a guideline for configuring their own systems.

Companies usually need more than one policy to cover IT infrastructure. They might need to create a cybersecurity infrastructure document as well, though it should not be separate from the IT infrastructure. Some companies actually don't have infrastructure blueprints in one comprehensive document. Instead, they break it up into a network diagram, a list of servers, a list and diagram of security controls and other separate documents. We think that the best approach is to get to know what documentation exists and then evaluate whether the company has blind spots that need better coverage.

Technical baselines will lay out configurations for IT services like Windows domain security, servers, workstations, or firewalls. There are a lot of external resources that can be used to help create these IT documents; the Center for Internet Security (CIS) is probably the most important.

BASELINE

Yes, that means a baseline isn't necessarily a document! A baseline can pretty well be a script with comments too! Also consider using collaboration tools that allow shared document editing to develop the policies that are on paper. Maybe you use GitHub and collaborate in Google Docs or use something similar. This is control, not anarchy, like some people mistakenly believe in the industry!

Templates should be used whenever possible, and setups should be automated and standardised. This way, the CSM can work with IT to control the policy in technical terms. The CSM should create the first baseline together with IT subject matter experts, enforce it, then start versioning it. A year or so later, review and improve it, then add some checks or security features into that policy before rolling it out again. Version control tools that are commonplace in software development should be used to maintain the version control and changes of these baselines.

CHANGE CONTROL

If there is one common denominator for all trouble in

IT and cybersecurity mishaps, it's uncontrolled change. Standardisation is key in IT. If there are no standards, then every machine is an individual snowflake, and the CSM cannot implement effective security. Standardisation enhances efficiency and manageability.

> **KEY TAKEAWAY**
>
> Only carry high-level policies that contain principles and requirements to be accepted by the top management. Make a clear distinction between the policy, baselines and their implementation in the policy.

So step one is to make IT policy manageable and efficient, then introduce security features. It can't be done the other way around, though it's tempting to set up security features on one machine at a time while leaving the others untouched. That's not going to work—there's no standardised management infrastructure in place.

IT directors may feel it's enough for them to say yes or no to a particular control; they get a warm, fuzzy feeling of being secure when they get to approve or reject a proposal. They feel comforted by sequential checkpoints that allow them to say yes or no to everything. But this is not the same thing as controlling changes in the network. Other people—not the IT directors—should be authorised to act on and enforce the baselines as quickly and efficiently as possible. Windows administrators should be able to roll out a change in security policy without management interference or having to wait for an official yes from the director.

Why should administrators be empowered this way, especially when there is a need for a very fast change? Consider what happens if there's a critical vulnerability, and they have to act within a few hours. Network outages are common. Maybe employees are unable to connect to the network or not able to use work-related applications, like email. It might be a partial or full outage or anything in between. It might include huge chunks of the whole internet being down. There's no way that any management process can keep up with these types of immediate needs within a short time frame. A policy should enable the work that a Windows administrator is doing within a certain time frame and that they are authorised to do.

ASSIGN THE POWER

Properly assigned empowerment of IT administrators and use of the right kinds of controls will make your IT infrastructure much more secure than any bureaucratic security decision-making process!

NETWORK SEGMENTATION

Now let's take a closer look at the solutions we recommend. Please note, some of the information we share in this chapter changes rapidly.

Let's start with network segmentation. This means the network is split into parts, and traffic between the segments is controlled. If someone breaches the network and there's no segmentation in place, the hacker will be able to access everything in the flat network. But if the network is segregated into segments, with routers and firewalls, the risk is distributed into smaller parts.

A variety of network profiles and security levels can be applied to segmentation. For example, there might be a general workstation network in one place and a production network somewhere else. There may also be an email and web server network called a demilitarised zone, or internet-connected servers networked somewhere. They're separated from each other by firewall rules so that not everything is accessible from one network to the others. That's great, but those rules can get complex.

A company that we used to work for had multiple countries in their internal networks—the United States, European countries and Japan. Their IT director told us that they had effective network segmentation in place, with limited access between countries. He said, 'These different countries cannot access each other.' So we connected to their network and tried to send simple echoes, ping packets, to the other networks. This is the simplest and first test anybody in IT would do to find out whether someone is online and connected.

If the network segmentation is effective, or it's behind filtering, there shouldn't be any response to the ping. But we did get a response. So then we tested some other types of network connections, actually trying to connect to the different systems and addresses in other countries. Everything—all of the services they were running in the different countries—was actually accessible. The IT manager's impression was not correct at all, and their network segmentation was not working as expected. The network segments were there all right, but there was no filtering of traffic between the segments.

Of course, while segmenting networks improves security, it also prevents people from using things, and there will be users who need access. There has to be a way to do that, so there need to be rules in place. The CSM's job is to figure out how to allow appropriate access as securely as possible. There are multiple ways to handle this need. First, identify the people who need to access the other segments. Second, select a solution that allows these people to connect to those segments flexibly and safely.

GATEWAY PROTECTION

Security often begins with a firewall—protection placed at the fortress gates, so to speak. Gateway protection might be the email filter that runs antivirus care and filters away spam. It might be a proxy server that protects the browser and browsing habits so that users are shielded from malicious content when they navigate to a not-so-safe page. Anything that performs perimeter checks by monitoring and controlling security measures at the border is gateway protection.

Gateway protection comes in many different forms, and it's usually managed by a network operations centre or security operations centre teams within the enterprise. It requires some kind of expertise to run these things—setting up email filtering services, web content management and proxy for the company, or an anti-spam and antivirus solution or a firewall. In fact, these controls might need a team to run them.

The trend these days is to buy such services from a cloud. An organisation might buy an antivirus or anti-spam solution for their email, or web filtering, all from an external service provider. In the past, companies used to do it by themselves. They installed a physical device and managed it themselves. Not anymore.

When buying these services from the cloud, companies should be aware that they're actually routing that traffic through that external provider, so they should be mindful of which partner they buy from because all of their email and web browsing traffic will be routed through that company and their assets. The Department of Defense in the United States won't buy these services from Kaspersky in Russia or any company in China, for example.

For most businesses, it doesn't make much difference what service they choose. They'd be happy just to have the service. It works, it's global, it's fast and it's efficient. It can be used from the office, from home, from anywhere—it just does its job.

CSMs just need to be aware that hiring a service creates a possibility for that vendor to monitor whatever goes through that solution. Maybe that's a good thing—it will allow the customer to view and monitor some aspects of the traffic, like security alerts or non-compliance. It's a great advancement and is now easier than ever to buy.

VULNERABILITY MANAGEMENT

A vulnerability is a security weakness in a system or service, usually one that can be exploited somehow. Exploitation means an attacker can try to take advantage of that weakness and may gain something in return—usually access to the system or information that he shouldn't have access to. Vulnerability assessment and management, as a practice, is a systematic way to find these vulnerabilities. After they've been identified, they have to be remediated and fixed. If a company is running a network of a hundred computers on the internet, for example, they need to be added to the vulnerability management process. Then they should be scanned periodically to find out if there are existing vulnerabilities that should be fixed.

Vulnerability management should not be done once—it should be more or less continuous. The CSM should give a strong message to the management that if the company doesn't manage vulnerabilities themselves, then someone with a darker agenda out there will do it, with potentially harmful consequences.

> Run vulnerability management as often as you can to catch any new weaknesses on time before attackers do!

Vulnerability management can be internal or external, and it can be passive or active. Active vulnerability management means scanning networks, sending out queries and packets, and finding out what systems, services and weaknesses are out there. Passive means putting something on your gateway to listen to all of the traffic that goes through, then deducing from that traffic whether there's a vulnerability in that service. A combination can be helpful.

> Companies would be wise to at least do vulnerability management on any of their external services that are visible from the internet. That's the bare minimum.

Let's say a company runs a website and ten more services that are available through the internet. It's quite easy to obtain a service from the cloud, buy from a consultancy, or do it themselves—but they should be scanning those services continuously. If any new vulnerabilities are found, they'll be alerted immediately and can fix them. Most CSMs are already doing these things or requiring them. If not, they're ignorant of obvious security risks.

Companies shouldn't forget that scanning the network with vulnerability solutions periodically has its limits. Most weaknesses are nowadays in web applications that are used with a web browser. Basically they are programs that run partially in the client's computer—and quite often, the client might be the attacker. Problems in web applications can't yet be reliably identified or fixed by automated scanner tools. Hence, we've been seeing a lot of high-profile data breaches when web applications get hacked—Facebook, for example, lost more than 50 million user accounts because they made a little mistake in one of their various applications. There are other approaches that try to remedy this weakness in scanner services. Some solutions install a small monitoring agent on the web application server itself, and this approach has the hope of detecting and even preventing attacks better than active scanners can. This approach isn't yet mainstream but somewhat complements the lacking features of scanners. No matter, a CSM would be wise to find web application weaknesses even if it means having to buy penetration testing against all of his internet-connected applications.

ACCESS MANAGEMENT

Companies should have a way to grant, change and remove access from people—usually, access to systems and information. A variety of options are available, but most companies today still use a centralised control called Microsoft Active Directory.

Active Directory has always been relatively easy to use: just make a change to Active Directory, reset the user password or create a user account, and everything would be there. Now it's getting more complex. We have mobile devices that need to be added to a mobile device management platform, maybe provided by Microsoft or a third party. There are Linux and Unix servers in the network that don't speak Microsoft, and web applications and web shops that are not integrated with the same credentials that people use at their computer.

At the same time, using this type of control is becoming easier because of web standards for authentication. Federation of identities between different web services ensures we can create identities within our network by using good old Microsoft Active Directory (AD) and synchronising with cloud services automatically. Companies use many different cloud services, like Microsoft Office 365, Gmail or G-Suite for Business. Federation allows them to use all of those services with the same credentials, which offers a centralised way to synchronise user access to third-party services in the cloud. Once a user is created to the AD domain, access to the company cloud service provider and more becomes a breeze, depending on how they integrate. Building this will take some effort at IT, but it will probably be worth it in the long run. Adoption of the cloud was just at its infancy in 2018, and we expect nearly everything to move to the cloud in the near future.

This offers a lot of efficiency because there's less to manage, but it's an enormous opportunity for hackers. Centralised access means someone can hack in, gain access to one of those cloud service providers or steal the password, then directly log in to all those services. There are no additional checks. It's a tradeoff that most organisations make gladly.

AUTHENTICATION

Authentication is related to access management, but it's a separate subject. Authentication covers the techniques and technologies companies use to ensure that people are who they claim to be. This is what makes it possible to have confidential information online. Without effective authentication, just about anybody could access a company's information.

To authenticate, we first have to identify the person trying to use the system. Typically, identification is when a user gives his username or his other identifier, like a mobile phone number. After the user announces his identity, authentication's job is to verify that the person is who he claims to be. In other words, the process begins when the user provides a bit of information that proves he is who he claims to be.

There are many levels and kinds of authentication, including password and two-factor authentication (2FA), which means another factor to verify the identity claim is required. It might be an SMS message to a registered cell phone number, a PIN token, a software certificate or something like that. It might be biometrics in some instances or a smart chip.

The level of authentication of identity depends on the company's needs. The best practice in authentication is to choose a level of security that's consistent with the values being protected. At a web bank where people are transferring money in and out, it might be a good idea to have two-factor authentication and various other security measures in place. This makes it inconvenient for people who use the application—they have to manage authentication devices, SMS tokens and those sorts of things—but it's worth the hassle for the added security. A web shop selling T-shirts, however, might not feel it's necessary to verify much about the users. They might not even care about the customer's identity, as long as they get paid.

> The biggest pitfall for authentication is when security people try to make everything ultrasecure without thinking about usability. It should be a balance of the two priorities.

The security people should understand a wide variety of different authentication technologies and methods to be able to pick the most suitable method for any situation rather than proceed with a one-solution-fits-all mindset.

In the past, companies tried to solve the problem of access management with something called a single sign-on (SSO) or related technologies. The goal of SSO is to implement everything behind one login so that users don't need to be concerned about more than one password for everything. This uber password is then stored somewhere and they hope it remains secure.

SSO is a good idea, and it sometimes works quite well. But companies should also remember that it creates a 'one password to rule them all' sort of risk. If people lose their one password that secures all their other passwords, they'll risk access to everything.

Psychological consistency is also important when dealing with access management. We've seen multiple cases where corporations have an enforced password policy in their Active Directory but totally ignore it elsewhere. In the best cases, AD password is mighty strong and synchronised and enforced across all workstations, servers and even some services in the cloud. Maybe it's a magnificent password with twenty characters. But, simultaneously, they're often offering internet services to their customers and allowing them to use any kinds of passwords. Quite often, their employees are using unintegrated third-party services or old systems that aren't consistent with their password requirements. What kind of a story are we telling to them about passwords as the CSM? When users look around and see what others are doing with their passwords, they see social proof that security management are okay with poor passwords across many places—but just not in the AD. The perception turns against security—it seems like a nuisance because of this inconsistency. They ask themselves, 'Why should I use a mighty hard password here when, everywhere else, it's so easy?' Try to be consistent.

ANTIVIRUS

Around the end of the 1980s, antivirus software emerged as one of the first computer software protections. Traditionally, antivirus software was based on signatures. The idea was to create a list of all malicious software in the world, then scan each computer at startup to see if there were any signs of any of them on the hard drive, stop the harmful programs, create an alarm and even remove them if possible. The obvious weakness of that approach is that it's impossible to maintain a comprehensive list of all the bad things in the world. Someone can always figure out how to create one more bad thing.

Today, antivirus companies have implemented something called heuristics and behavioural modelling—even reputational modelling. It sounds fancy, and it works to an extent. Of course, the hacker has the benefit of figuring out one more bad thing that works differently, and they do. It's a cat-and-mouse game, where the mouse is always a few steps ahead.

Any experienced security manager knows that they shouldn't rely on that antivirus software too much. We've done a lot of tests against antivirus penetration, which means we create malware that's supposed to go through and install undetected. Our success rate is high: 96 to 98 percent of the time, we can bypass the antivirus protection. If we can do it and we aren't virus writers by profession, almost anybody can do it. In practice, all a knowledgeable programmer needs is a text editor and a compiler, and maybe half an hour of time, and he will be able to modify almost any malware variant to be undetectable by almost any antivirus program.

Having acknowledged that antivirus approaches have always had their weaknesses and flaws, we still think it's necessary to run one. If you get ten thousand attempts, antivirus will block a major portion of them. Antivirus software might find 920 identifiable incidents out of one thousand, leaving only eighty infections undetected. Antivirus is not 100 percent secure and never will be, but it can cut down the numbers. It's what doesn't get caught by the antivirus that you should be hunting for. This is why we noted earlier that companies would be wise to consider workstation networks already compromised, even hostile, to the other networks. Consider the previous example of eighty undetected infections—we don't know anyone in IT or cybersecurity who's comfortable with this fact. Numbers may be different, but the basic idea is a hard, indisputable fact.

BEYOND ANTIVIRUS

Some people sell solutions that claim to be able to identify malware that antivirus can't, like sophisticated hacking attempts directed toward high-value targets. Take these claims with a grain of salt because it takes a bit of effort to buy, manage and implement them, and there's still very little evidence that they deliver the value promised. It might be a good idea to try out these services on a demo or a free trial. But the question remains, how to verify their claims? They're looking for an invisible threat that antivirus doesn't detect. There's no way to be sure they're doing anything at all; even if they provide a sample virus, it's probably a sample they know will be caught. Some approaches involve sandboxing applications or portions of the computer to secured envelopes that are monitored, and many attacks are stopped on the fly. The approach has shown promise but isn't actually handling the real problems—the gullibility of people and the security flaws in the system and applications themselves. Nevertheless, this approach can be an additional layer of defence and can probably stop some attacks that would be successful otherwise.

CLOUD FIRST BUT NOT LAST

If you want to be efficient, get your first layer of malware and spam protection from the cloud. Then do the same checks but with different products and possibly even with a different approach at the workstation level. This shouldn't be overly expensive and provides a fair level of protection against simple attacks.

LAYERED ANTI-SPAM AND ANTI-MALWARE

Spam filtering is a productive technology. With no spam filtering, life quickly becomes horrible. People using spam filters save time and effort, and are less annoyed by computers. Spam filters are a boon to usability, especially when combined with anti-malware, which scans emails for any threats. Some companies even add two layers—first an anti-spam and anti-malware to pre-filter everything at the servers that process the messages, then another malware scan at the workstation afterwards. This last check isn't uncommon and might add a little bit of security, though usually not much.

To find out how much, some services let companies and hackers test their malware against all kinds of antivirus products. The basic idea is that anyone can go to a website and upload their malware variant, then run an automated test against all the existing antivirus engines and get the reports. How many of them caught the malware? How many layers of antivirus are secure enough for the organisation? Or is any of them able to detect the threat? What these tests reveal is that running the same kinds of security checks one after the other won't add much security. If malware passes a few of the antivirus products, it often passes the rest too. It's like having a door with a certain key—it doesn't make sense to have three more doors just like it, all with the same key. It doesn't add security. There should be a different key or a different kind of door.

SECURITY EVENT COLLECTION AND LOG MANAGEMENT

In an ideal world, companies would have a centralised service where they store all of the event logs of all of their systems, including anything related to security. All the entries would be pushed or pulled into one centralised location, where they would be data mined, correlated and stored efficiently. Then, out of that big mass of data, machines would automatically detect anomalies with fancy algorithms. That's the dream.

Of course, this is easier said than done. The CSM will have to answer a lot of questions before deciding on the right solution. What if it's a global company with twelve or more time zones? All of the events across the globe are pushed to this central system, so what time is marked when the entry arrives? If the company is running its own time service, what if it's a couple of hours off? Can these events be correlated afterwards? Do the systems provide a way to log entries or transfer those logs away from those systems? How are those logging features enabled? Which systems should be covered?

Event collection requires recording events in a log, then transferring them somewhere. If someone disables the log feature on a system or silently removes some entries, the centralised log management system won't be good for detecting anything on that system anymore. There will be no meaningful events to transfer. Or if an attacker gains access to the system by a means that's not actually logged at all—which is usually the case—it wouldn't be apparent in the centralised system, even if all logging is enabled. A lot of cyberattacks and exploitation techniques leave no log entries because they don't require running software in the system that generates logs. After the exploitation, the hacker has access to the system and can disable the logging, change the logs, remove entries or add false entries. Even in the case that some logs are generated, the security team will most likely miss it because of 'alert fatigue'. They deal with thousands of alerts on a weekly or monthly basis. This desensitises them against any real alerts.

> Log management is a good tool for investigating all kinds of problems and security incidents but also, even more importantly, almost any problem at IT in general.

Companies that buy log management solutions have to trust that they will enable the company to analyse security incidents after they happen as well as detect if something is happening. In reality, some things will be identifiable, but not most; a skilful hacker doesn't create a log entry. Antivirus won't detect the hack, the logs won't say anything, and yet the system will be compromised.

Security shouldn't be the only reason for acquiring one of these event collection and log management systems. It should be purchased for manageability and problem resolution in IT, but the business benefit should be the main selling point, not the security benefit. These event collection and log management systems don't make the greatest difference in security.

SECURITY INFORMATION AND EVENT MANAGEMENT

Log management can evolve into security information and event management. With log entries or threat information entries and sources in the network listed and stored in one place, companies can correlate that information with wisdom and artificial intelligence. That's security information and event management; it sounds powerful, and it can be.

These systems make sense for large organisations that run security operations centres. Service providers that sell cybersecurity services and continuous services to other companies have good reason to invest in one of these. They could integrate all of the different sources of information into this one system—vulnerability management and network scanning, firewall traffic and intrusion detection. Quite often, companies subscribe to external threat feeds as well, which provide external threat intelligence. This would give them as much information as possible from different sources, then let the intelligent machine decide which of those alerts constitutes a risk.

For many security information and event management (SIEM) solutions, the promise is like this: first, there's a hacker on the internet scanning the company's network. The SIEM solution would identify a network scan on the network edge based on the firewall log entry. SIEM could also use external threat feed data and notice that the attacker's IP address belongs to a known bad actor. It would be possible to raise an alarm just with this information at hand. Usually, the hacker would launch an exploit on a found vulnerability on a server, and the network intrusion detection system would log an entry to SIEM that there was an exploitation attempt that it saw. This would be logged, maybe with a higher priority. So first there was a scan from a known bad actor, and now there's an exploitation attempt.

Then, staff could go back to the database in the SIEM solution and cross-check whether the system in question was vulnerable and whether there was a vulnerability that would match that exploit. Or this step could be fully automated based on the capability of the SIEM solution. If so, then the incident would be given a very high priority—

because there was a vulnerability that was exploited and was probably successful. There would be a pretty good chance that the attack was successful.

This story is fine, and sometimes things will play out like this, just like they are told in SIEM vendor presentations. In reality, there are multiple factors that mess up that pretty picture.

The usual signal-to-noise ratio between real alerts and false positives in SIEM systems is typically so high that the systems are often unable to deliver the value that vendors promise. Quite often, real attacks are lost like a needle in a haystack. In most cases, the systems aren't able to do the correlation with the precision and efficiency that the vendors promise.

Lots of false alerts means there has to be an expensive team managing them very actively, all the time. The data sources need a lot of pruning, rules need to be modified and created and so on. If you do not have the resources to run something like that, don't buy it. If you plan to run a security operations centre (SOC), please bear in mind that you may have to pay a steep price for a meagre return.

THREAT INTELLIGENCE SOLUTIONS

SIEM and threat intelligence solutions go hand in

hand nicely. An internal SIEM solution would only be able to correlate events that are internal to the company's infrastructure and any traffic that comes that way. Many companies subscribe to a number of threat intelligence feeds. These offerings generally fall into two distinct categories:

1. Cyber intelligence solutions
2. Threat feed solutions

A CSM should be able to distinguish between these. A threat feed contains something called indicators of compromise, or IoCs, which are basically technical fingerprints of known bad actors or their techniques, tactics and procedures. A typical IoC could be a hash value of a known malicious file, or the IP address belonging to a known bad actor. These feeds can enrich the internal SIEM by adding an external view to what the system is able to correlate. The drawback is that the signal-to-noise ratio still usually tends to be quite high, and SOC teams will see a lot of false positive alerts.

Cyber intelligence solutions can also be integrated to a SIEM, but with the distinction that their data is usually enriched by analysis before it is sent to the customer for consumption. Hence, these solutions have a naturally low rate of false positives, and the information they send tends to be more reliable than mere technical IoC feed data. Basically, a good cyber intelligence solution can always provide actionable alerts to the customer without the need for further correlation with other events in the SIEM before SOC analysts are able to act on the information. This provides a remedy to the alert fatigue problem that is plaguing the cybersecurity industry.

Think of it this way—all actionable cyber intelligence needs to be collected and analysed before it becomes consumable and actionable—before anyone can do anything about it. A CSM would be wise to externalise some of that burden to a service provider if he can't have it done properly in-house. Most companies don't have that capability and will not want to invest in it.

INTRUSION DETECTION SYSTEMS

Intrusion detection systems (IDSs) are an old idea, in theory very similar to antivirus. The concept is that an alert should be created if there's an attack flying over network traffic that has an identifiable fingerprint or behaving in a way that's suspicious. It's based on the idea that there's a certain number of bad things that can be identified on the fly and that you can make a list of them all. Like antivirus, it's nearly impossible to list everything; something will be missed. Likewise, the behavioural approaches have the typical problem that they tend not to be 100 percent effective and also increase the alert fatigue that SOC teams are too burdened with anyway.

IDSs come in two variants: network-based and host-based. Network-based IDSs (NIDSs) listen to network traffic, usually at the choke point, like one side of a firewall. The other variant, host-based IDSs (HIDSs), is the software or rules that are applied to a single system, like a server. NIDSs listen to network traffic and give alarms, while HIDSs monitor the integrity of the system and any bad signs of something happening inside the system itself.

ASSIGNING ANALYSIS

A core question in cyber threat and intelligence services is this: who's doing the analysis phase before information can be consumed? This will tell you who's going to have to employ the staff to do it. This takes a large budget that most companies can't bear. If this is your case, go for an intelligence solution instead of increasing your own alert fatigue.

The focus of HIDSs is usually to monitor any important changes within a system. Let's say a company is running a Unix server. A HIDS running in the server notices that binaries, or executable files, are changing. An alert would be raised that a change is detected, but not necessarily where it came from. Maybe someone internally made the change by updating the binaries to a newer version, or maybe a hacker signed in and is replacing executables with malicious ones.

The obvious pitfall of this technology is that it creates a lot of false alarms. There's also a high chance that any attack will fail to be detected at all. In practice, these detection systems work like an alarm bell that doesn't always ring, and when it does, there's usually nothing to worry about. Is it really a good alarm then? If it were a fire alarm, you'd probably throw it away after a week. The technology isn't necessarily bad, but it has its limitations and requires a lot of pruning and maintenance.

BACKUPS

We've talked a lot about backups already, and our advice is simple: do them. Do them first, test them, and do them again. Have a process in place that ensures they actually work. If you don't do anything else in security, do this. Don't even have a firewall if you don't have the money, but have backups. Take the backups offline at least once in a while.

Online backup has become very popular. Everybody has an iPhone or Android, and if you take a photo, after a minute, it will be in the cloud, backed up. Companies should look for similar solutions that are equally easy for the user—automatic data transfer, low maintenance and an adequate level of security. It might not go to Google's or Apple's cloud, but there are separate solutions that give you the promise of online backup with encryption and security and central management. Companies should look into this because almost everyone has already adopted the idea of online backups based on their personal experience with mobile devices. There's no adoption curve.

A good online backup service should provide you with the innate ability to have data offsite and available immediately for restore when it's needed. Many services offer encryption, user management and other security features that should make it easy for most businesses to adopt these services. They also provide some protection against ransomware attacks that are commonplace nowadays.

VULNERABILITY BLIND SPOTS

We are constantly surprised at how organisations fail to recognise their security blind spots. Sometimes, these blind spots are technical in nature; other times, they are failures in physical security or access control. Here's a real-world example.

We worked with a major company in Europe that's in the real estate and facilities construction business. They are one of those entities who are considered critical infrastructure for that country. They had virtually hundreds of physical locations and huge buildings. Most of their access control systems are online, connected to the internet.

The servers that controlled the doors were grey metallic boxes that were screwed onto the walls in a dusty room and essentially forgotten about. This company had these in most of their buildings, usually somewhere locked away in the basement floor. The boxes were connected to the internet by routers that were managed by the service provider who had installed the system. Interestingly, but not surprisingly, most of the routers had been installed in 1999 and were actually managed by nobody. It's very commonplace that these boxes quickly become forgotten by everyone because all they care about is that access to the buildings works and nothing more. We inspected some of these routers and found that they were full of multiple vulnerabilities that could have been exploited from outside of the network. Anyone from the internet could find the system and scan it and see that it was vulnerable to all kinds of known exploits.

Of course, someone did exploit that modem vulnerability, gained access to the company's built-in access control systems and used them to their advantage. Remarkably, before the attack, IT didn't have any idea that this was a risk.

The worst part was that even after they were exploited, they still didn't consider it a serious problem. This was surprising, especially considering they had a law enforcement agency in one of their facilities. Despite all that, they probably still have the same vulnerable modems in use today.

The point of this story is that even the biggest and most forward-thinking companies in the world have security blind spots and vulnerabilities that they don't know about. Or they do know about them but fail to fix the problem. That's why it's important for every CSM to follow the steps and guidelines spelled out in this book. By adhering to the suggestions in these chapters, security blind spots can be uncovered, and they can be addressed. Ideally, before a breach or an attack happens.

CONCLUSION

We began this book considering the situation of a company that wants to hire a CSM, and of the CSM who is looking to be hired. Throughout, we've explored key aspects of that partnership. Like any relationship, the relationship between employer and CSM begins with a combination of optimism and naivety. Neither side knows very much about the other but both hope that the other will meet their needs.

For a company hiring a new CSM, it's hard to tell whether their new employee is a good fit for the role. Even if the interview process is rigorous, will their new hire be a good cultural fit? For the new CSM, there's a danger that the company doesn't fully understand the risks he has been hired to mitigate. If they did, would they need to hire a CSM?

The journey of understanding doesn't conclude with the signing of contracts. It starts there. Both the CSM and the company need to make an effort to understand the other's priorities and expectations. The CSM has a huge job ahead to learn how the business actually works, and the company has the task of understanding their cyber risks. Both parties need to communicate accurately and transparently.

In the best case, the company's expectations and the abilities of the CSM will fit together beautifully and a successful cybersecurity partnership will ensue. In the worst case, poor communication and mismatched priorities can lead to disaster.

For this reason, it's essential that CSMs buck the stereotype of the geeky, socially isolated techie. A good CSM must have excellent communication skills; otherwise, a disconnect between their understanding of their role and the company's will open up and continue to grow. The role of a CSM may appear to be a technical one, but there's actually a strong people element to the management of risk.

Risk can be hard to define, as can organisational politics. A CSM entering a new role will likely encounter both formal and informal leaders, all of whom have worries. Some of those worries will be quantifiable; others less so.

In all likelihood, each leader will imagine that the risks they face are the most pressing, without an objective understanding of how large those risks are in the context of the entire company.

How should the CSM approach this situation? Should they simply tackle the risks they deem most important? That's a recipe for panicky leaders and negative feedback. A CSM must acknowledge the asymmetry of risk understanding, and the concerns of stakeholders within the company, then find a way to quantify those concerns. If he can do that, he should expect that the business will give him the resources to address them. Typically, he won't have access to the resources to fix everything within his department's budget, so he must rely on other departments allocating a portion of their budgets to addressing cybersecurity. When everyone works together, all parties get their needs met. That's the foundation of a good partnership.

Ideally, both parties will come to know each other well. The CSM will recognise the structure of the company, key people and their motivations. The company, meanwhile, will make an effort to understand the needs and desires of the CSM and provide necessary support. Together, both the company and the CSM can plan for improvements, share successes and address challenges.

Hiring a new CSM is a harbinger of change. If that weren't

the case, there would be no need to make an appointment. Some people will welcome change, others may be indifferent and still others will actively resist. Engaging skilfully with all these different groups requires a high level of empathy and organisation, not merely technical skill.

Additionally, a CSM must communicate with people from a broad range of departments, who may perceive the value of cybersecurity very differently. The management team has the overall responsibility for handling risk in the company. The risk management unit looks to understand the business angle of any risk and find ways to manage the downside. The finance team takes an overview of the company's financial position and allocates budgets. Maintaining good relations with the finance team is an essential part of any CSM's job.

Other units include HR, education and facilities. It's important that employees throughout the business receive education about the value of cybersecurity. Facilities may not be the most glamorous department, but it controls physical cybersecurity risks, such as securing server rooms.

People in operational business units—sometimes called production units—may be focused entirely on the day-to-day roles that bring revenue into the company.

Nonetheless, they need to be aware of the continuity risks that occur when processes change. These units may be largely independent from other areas of the business, so it can be valuable to employ a cybersecurity champion to communicate with them about risks and benefits.

Another unit is the legal department, which controls the wording of contracts. Whenever one company engages with another, a contract—usually including security clauses—will be involved. It's important that these clauses match the cybersecurity needs of the businesses involved. At minimum, they usually involve a non-disclosure agreement.

Last but not least, IT plays perhaps the most crucial role in the successful implementation of cybersecurity. Most technical cybersecurity controls, from basics such as backups and firewalls to the most sophisticated technical solutions, are implemented by the IT department. Therefore, it's crucial for a CSM to win the IT department's buy-in.

In part III of this book, we discussed an ambitious goal: building a lasting cybersecurity architecture in approximately ninety days. This may seem like a short period of time, especially for organisations starting from scratch, but it can be done.

This part of the book focused on the elements that

typically need to be addressed in any company that is contemplating a major cybersecurity overhaul. The process begins with agreement about cybersecurity policies and procedures. Who needs to agree? At the very least, CSMs and leaders. This, of course, depends on those changes being beneficial to the company as a whole.

Next, there will be various controls that define what needs to be done. These may be policies, procedures or technical security technologies. The company and the CSM will need to address questions such as controlling access to different systems, risk management, compliance and assurance, continuity management and disaster recovery. There will be questions about IT-related security controls, security leadership, how to be a good leader and how to work with HR.

Each one of these elements requires the CSM and the company to work in partnership. If communication is strong and open, every conversation will become an opportunity to strengthen that partnership, deepening understanding and refining shared goals, and moving toward those goals. If communication falls apart, the partnership will soon erode and ultimately collapse.

At the outset of this book, we asked ourselves, when is the best time to buy fire insurance: before or after your house burns down? That's a rhetorical question. The answer is obvious. By the same logic, when is the best time to invest in cybersecurity: before or after your company gets hacked?

Unfortunately, most companies and organisations invest in cybersecurity or hire a cybersecurity manager only after they've experienced a data breach and seen the tremendous damage it caused.

If an organisation wants to be safe from cyberattacks, they have to care about security. They have to prioritise it. They must realise that it's not a discretionary expense. This means hiring a CSM and then allocating budget funds to pay for what needs to be done.

Cybersecurity managers are agents of change who have the power to transform companies for the better. Cybersecurity isn't about computers, servers, firewalls and software. It's first and foremost about human behaviour. It's everything we do with computers and smart devices; it's about the passwords people choose. It's about their use of Facebook, Skype, Dropbox, cloud services and so much more.

Engaging with people and making the case for a positive shift in human behaviour is a big part of the CSM's job. A good CSM sees themselves not only as a technical expert but an advocate for keeping people safe online.

Our ambition is for companies to treat cybersecurity as a people problem. Cybersecurity is only as good as the people who use it. When this shift occurs, and we think of cybersecurity not merely as a technical issue but as a people issue, we will all live in a safer world!